Image Transfer on Clay

Image Transfer on Clay

Screen, Relief, Decal & Monoprint Techniques

Paul Andrew Wandless

LARK CRAFTS

Asheville

editor: Suzanne J.E. Tourtillott

art director: Kathleen Holmes

cover designer: Barbara Zaretsky

associate editor: Nathalie Mornu

associate art director: Shannon Yokeley

art production assistant: Jeff Hamilton

editorial assistance: Dawn Dillingham,
Delores Gosnell

editorial intern: Megan Taylor Cox

art intern: Ardyce E. Alspach

photographer: Lisa Goodman

Page 3, clockwise from top left: Amanda J. Wolf, *Blue Landscape Teapot with Bird in Flight*, 2004; Paul Andrew Wandless, *Good Boy*, 2005; Amy Sanders, *Cut Vase*, 2004; Scott Bennett, *Saucer*, 2005

PEPSI and the Globe design are registered trademarks of PepsiCo, Inc. Used with permission. Trademarks of PepsiCo, Inc. used in this publication are for illustrative purposes only. Such use does not imply an endorsement of the material.

An Imprint of Sterling Publishing
387 Park Avenue South
New York, NY 10016

If you have questions or comments about this book, please visit: larkcrafts.com

The Library of Congress has cataloged the hardcover edition as follows:
Wandless, Paul Andrew.
    Image transfer on clay : screen, relief, decal & monoprint techniques / Paul Andrew Wandless.
        p.   cm.
    Includes index.
    ISBN 1-57990-635-4 (hardcover)
    1. Handicraft.   2. Transfer-printing.   3. Clay.   I.   Title.
TT880.W365   2006
738.1'5—dc22

2006006725

10 9 8 7 6 5 4 3 2 1

Published by Lark Crafts
An Imprint of Sterling Publishing Co., Inc.
387 Park Avenue South, New York, NY 10016

First Paperback Edition 2013
Text © 2006, Paul Andrew Wandless
Photography © 2006, Lark Crafts, an Imprint of Sterling Publishing Co., Inc., unless otherwise specified

Distributed in Canada by Sterling Publishing,
c/o Canadian Manda Group, 165 Dufferin Street
Toronto, Ontario, Canada M6K 3H6

Distributed in the United Kingdom by GMC Distribution Services,
Castle Place, 166 High Street, Lewes, East Sussex, England BN7 1XU

Distributed in Australia by Capricorn Link (Australia) Pty Ltd.,
P.O. Box 704, Windsor, NSW 2756 Australia

Manufactured in China

ISBN 13: 978-1-57990-635-1 (hardcover) 978-1-4547-0332-7 (paperback)

For information about custom editions, special sales, and premium and corporate purchases, please contact Sterling Special Sales Department at 800-805-5489 or specialsales@sterlingpub.com.

Requests for information about desk and examination copies available to college and university professors must be submitted to academic@larkbooks.com. Our complete policy can be found at www.larkcrafts.com.

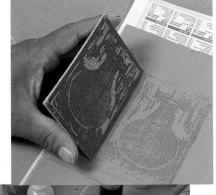

Contents

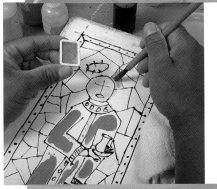

clay and printmaking

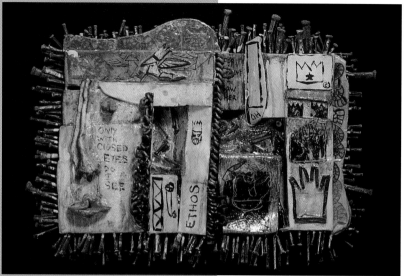

Paul Andrew Wandless
Ethos, 1997
11 x 7¼ x 2 inches (27.9 x 18.5 x 5.1 cm)
Hand-built low-fire clay; slips, nails;
collaged screen prints
Photo © artist

Ways of transferring an image from one surface to another have been around for thousands of years and the techniques that have been developed to do so are simple, easy, and repeatable. Printmakers create dynamic images by incorporating drawing, painting, and photography with graphic methods such as screening, mono-printing, and relief. Using such techniques to put images on clay seems natural to me. Clay, after all, is the creative medium with the most receptive surface. The soft material is like a blank sheet of paper or a canvas on which you can use printmaking techniques to express ideas using color, narrative, symbols, marks, and textures.

My journey as an artist—searching for ways to manifest my creative ideas via the most appropriate vehicle—has taken me down many paths. Mostly I use clay, but I also love drawing, sketching, painting, printmaking, photography, and digital imagery. I feel it's important to explore all your interests and to choose the techniques and mediums that best suit the aesthetic demands of the image. Faithful to this philosophy, I'm always looking for ways to weave together my creative passions and get the most out of what they have to offer visually. Image transfer techniques for clay using printmaking processes have opened new creative doors for me. My first image transfer onto clay was a relief print of twine on a hand-built platter. Early works used my printed images on paper to apply collage on the clay surface, but I knew there was a way to better integrate the two mediums. My mentor, artist Victor Spinski, has since taught me much more about plaster, overglaze decals, and screen printing. My printmaking forays and interactions with still other print-making and ceramic artists help me continue to experiment and to dis-cover even more transfer methods that I can adapt to my work.

Paul Andrew Wandless
Green vs. Orange, 2005
11 x 13½ x 1 inches (27.9 x 34.3 x 2.5 cm)
Low-fire clay; underglaze, majolica; clay monotype from plaster slab
Photo © artist

Many artists who use image transfer techniques were generous with their time, advice, and tips as this book was being written. You'll see much of their work in the gallery images they contributed. These ceramists all have the same passion to explore combinations of creative mediums and learn new methods. Histories of clay and printmaking and in-depth descriptions of transfer techniques have been published before now, but this book breaks new ground by illustrating the processes step-by-step, with photographs and finished examples.

Image Transfer on Clay is an introduction to basic image transfer techniques for the home studio. Since there are so many ways to transfer an image, I decided to focus on basic methods that you can explore and adapt to your own work. One common misconception is that you should be an experienced print-maker to be able to use transfer techniques. Truth is, the basic methods of screening, monoprinting, stenciling, stamping, relief printing, and making decals are very simple. Information on how to make your own screen and a plaster slab are also covered here, so you can truly start from the beginning of the process. Victor Spinski demonstrates how to make a two-color image with hand-mixed overglaze decals. Then I show how to make a multicolor decal using premixed overglazes.

Paul Andrew Wandless
Little Boy Blu, 2004
13 x 12½ x 1 inches (33 x 31.8 x 2.5 cm)
Low-fire clay; underglaze, majolica; clay monotype from plaster slab
Photo © artist

This book will introduce these techniques with easy-to-use processes, easy-to-find materials, and familiar ceramic tools and equipment. Once you learn the basics, you'll be able to explore and experiment on your own, customizing the processes for your own work. My goal has been to put all of the information you need to get started here in one resource, and that it be both easy to understand and simple to execute. I hope you become as excited as I am about what is creatively possible once you've been introduced to image transfer techniques for clay.

a brief history of printing on clay

Hope Veronica Clare Leighton
Decal from New England Industries Series Depicting Farming, 1952
10 inches (25.4 cm) in diameter
Image courtesy of the Wedgwood Museum Trust, Staffordshire, England

Cylinder seal from Susa, Iran, decorated with drinking cattle and inscription, 3rd millenium B.C.E.
Height, 1 inch (2.5 cm)
Steatite
Photo by Gerard Blot
Collection of Louvre, Paris, France. Réunion des Musées Nationaux/Art Resource, NY

Clay and printmaking have a long history together. Ceramic artists have been able to benefit creatively from the marriage of these two dynamic mediums by using the simple strategy of replacing sheets of paper with slabs of clay and printing images onto clay with ceramic colorants instead of inks. Some printmaking techniques are better adapted to clay than others, but all of them can be mined in some form or fashion.

Early examples of stamping, embossing, and stenciling are evident on ancient shards of pottery and in tile- and brick-clad architecture. Today, computers, joined with new ceramic technologies, assure the continued relationship between clay and printmaking—with even more avenues of exploration awaiting us.

Stamping (also called *relief printing*) is one of the oldest types of transfer methods used in ceramics. It's been used by potters and sculptors alike to sign and create images on their works for thousands of years (see photo at left). Images were impressed on soft clay with carved stamps and cylinders as early as 4000 B.C.E. by the Sumerians of southern Mesopotamia (now Kuwait and northern Saudi Arabia). Two millenia ago, Egyptians carved emblematic stamps made of hardened clay or stone. They were used to affix a seal of wet clay to a container or door; when dry, the seal acted as a kind of ancient alarm system. Ceramists still use hand-carved and

The Poultry Women. A mid-19th century polychrome underglaze transfer print by F & R Pratt of Fenton, Staffordshire. Original drawing by Jesse Austin. 7¼ inches (17.8 cm) in diameter
Courtesy of Brenda Pegrum

Pearlware Jug, circa 1813. Photo shows both sides.
Courtesy of Garry Atkins

manufactured stamps made in all shapes, sizes, and materials (see photo at bottom right).

Stenciling is another image transfer technique that dates back thousands of years; most likely it is the ancestor of screen printing. Stenciling made its first appearance in early 2500 B.C.E. in Egypt and Greece, and in the first and second centuries A.D. in Japan and China. During the 17th and 18th centuries, stenciling was used for decorative purposes in England and France, on such surfaces such as wallpaper, furniture, textiles—even floors.

Nearly every culture has employed some form of the stenciling technique. They've made creative use of a variety of thin materials, including banana leaves and mulberry bark early in the history of the Polynesian Islands' culture. Designs were cut into the organic material and dyes forced through the openings, transferring the images onto bark cloth, or *tapa*. Today a wide variety of commercial stencils is available to clay artists for transfer images. Ceramists often make their own custom shapes, though; they're so easy to make by hand.

A more detailed use of the stencil was created in early 18th-century Japan. A detailed design was cut from two sheets of waterproof paper, and threads of fine human hair were glued between them to hold the thinnest and most intricate parts

Permoplast Modeling Clay, produced by American Art Clay Company, Inc. circa 1930–1950s
Photo © AMACO

Death of Wolfe Teapot, circa 1775
Earthenware; transfer print engraved by
William Woollett
6½ inches (16.5 cm) tall
Image courtesy of the Wedgwood Museum Trust,
Staffordshire, England

Grimwades Ltd.
Pudding Basin, 1911
Height, 4¼ inches (10.9 cm)
Glazed ceramic
Photo © Museum of London

George Cartlidge
Abraham Lincoln, circa 1920s
6 x 9 inches (15.2 x 22.9 cm)
White clay, glazed
Photo © AMACO

of the image in place. This may be one of the earliest applications of the screen-printing principle at work. Variations of this type of the screened stencil were used in different parts of the world for simple image transfers. As trade and travel between Japan and the Western world began, this more sophisticated form of stenciling developed outside of China and Japan.

The 20th century saw rapid improvements in screen-printing technology. Screens made of silk and synthetic fabrics were stretched onto wooden or metal frames; registration techniques made multicolor screen printing possible. Stencil film, developed in 1929, could be cut with a knife and glued to a stretched screen, making intricate images much easier for the commercial market. Joseph Ulano later developed an improved stencil film that is still being used today. The stencil material soon stimulated artists like Pablo Picasso to experiment with it.

During the 1930s a wax-and-glue method was devised. An image was first drawn onto the screen with a waxy crayon, then the whole screen was coated with a glue solution. Once the glue dried a solvent was used to dissolve the wax, creating open areas that the printing inks could pass through. The same basic procedure, with improved materials, is still used today. At first, modern screening methods could only be used to reproduce hand-drawn and hand-cut images. Photographs, woodcuts, engravings, drawings, and paintings were reproduced onto clay. English transfer prints from the late 18th and early 19th centuries (see photo at top left) illustrate the degree of detail that was possible with the technology of the day. Commercial ceramic ware, rich with advertising images and text, may well be some of the most familiar joining of printmaking and clay (see photo at top right).

It wasn't possible to transfer photographic images onto ceramics until a light-sensitive emulsion was invented that could take the place of the glue screen filler (see photo at bottom left, page 10). With this breakthrough, the artist became free to explore any and all imagery—screen printing no longer has any real image limitations.

Ceramists soon started screening photographic images onto decals with overglazes; the process really took off in the '60s and '70s, and decals are definitely popular in contemporary clay work (see photo at top right). Commercial decals have long been screened for use in industry and production pottery, but they're also used as collage material by many ceramists. Screen printing is even more popular with ceramic artists today: it's used to transfer complex images onto decals, onto plaster slabs for monoprinting, and even directly onto the clay surface itself (see photo at bottom right).

Technology continues to be the driving force in new image transfer techniques. Newer image manipulation software, computers, printers, ceramic materials, and decal papers are developed every year. Computers let the artist create, copy, resize, and change images in ways never possible before, and they can do it more quickly, with greater sophistication, than ever. The darkroom, copy machines, and other complementary forms of creating or manipulating an image are being replaced by these latest technologies. New and improved commercial materials for ceramic image transfer are now more accessible, environmentally safer, and more cost effective. Combining clay and printmaking techniques has never been easier or more exciting than it is today.

Juan Granados
Collection: Abuela, 2005
14 x 14 x 3 inches (35.6 x 35.6 x 7.6 cm)
Wheel-thrown whiteware/low-fire clay; electric fired, cone 04; photo image transfer
Photo © Wesley Harvey

Howard Kottler
Look Alikes, from the set of four dinner plates, *American Gothic*, 1972
10¼ x 1⅛ inches. (26 x 2.9 cm)
Porcelain, decal
Photo by Paul Macapia
Seattle Art Museum, Gift of the Howard Kottler Testamentary Trust

screen prints

Sharon B. Murray
Decay, 2004
10 x 9½ x 8 inches (25.4 x 24.1 x 20.3 cm)
Slab-built stoneware; electric fired, cone 07; silk-screened decal, cone 07;
china paint, cone 017
Photo © Greg Staley

Screen printing is a popular method for transferring an image onto just about any surface. It's been particularly popular for clay artists, partly because the needed materials are more accessible and reliable than ever. There are a variety of simple methods for putting an image onto the easy-to-make-and-use portable screen, which you then use to print the image onto any ceramic-friendly surface, whether a decal, a plaster slab—or even directly onto the clay itself.

Instead of ink, an oil-base overglaze is used to screen print the image. All the materials that were used in the demonstrations in this book were chosen because they're relatively safe, but make it a practice to always read the manufacturer's recommendations, too.

Commercial screens in standard sizes may be purchased, and a mail-order service can apply the image for you, but it's simple and inexpensive enough to do it yourself. That way you'll have greater freedom over its design and content (just remember to honor all copyright and intellectual-property rights if you "appropriate" another artist's image). Your regular studio space is a fine place to work, and common hand tools and inexpensive materials are all that are needed to make the screen and print the image. Constructing a reusable screen takes less than an hour to complete.

Richard Burkett
Feast and Famine Platter, 2005
1¼ x 14 x 6½ inches (3.2 x 35.6 x 16.5 cm)
Slip-cast porcelain; gas fired in reduction, cone
10; laser printer decals, cone 05
Photo © artist

Jeff Irwin
Fear and Herding, 2005
17 x 17 x 1 inches (43.2 x 43.2 x 2.5 cm)
Earthenware; electric fired, cone 04; laser toner
glue transfer, laser toner decal, cone 04 glaze,
scratched, wet work
Photo © artist

Making a Screen

The screen used for printing is typically a fine-mesh woven material, stretched and stapled over a wooden frame. You can purchase preassembled commercial screens, but making your own custom-size screen requires only common hand tools, wood, and screen fabric.

Buy kiln-dried frame sections (sometimes called *stretchers*) from an art store or screen-print supply store. They're durable, resist warping, and come in different lengths, ready to assemble right away. Such stores aren't always close by, so another option is to purchase wood from a local lumber store. Clear (knot-free) pine, spruce, or poplar are the best types of wood if you want a durable frame. I used mitered pieces of 1 x 2-inch (2.5 x 5 cm) pine for the demonstrations in this book.

Years ago, screen fabric was woven from fine silk threads, but nowadays it's

Photo emulsion screen with digital image, overglaze in tubes, and squeegee

made from either multifilament or monofilament synthetic materials. Multifilament has many fibers twisted together to make the strand. A monofilament fabric consists of an individual smooth filament, and this type of screen produces a sharper image than the multifilament type.

An array of screens with images made by various methods described in this chapter

Brendan Tang
Shooting the Messenger, 2005
12¼ x 8 x 6 inches (31.1 x 20.3 x 15.2 cm)
Wheel-thrown stoneware; electric fired, cone 6;
commercial decals, luster, 018; Lazertran, paint
Photo © artist

Polyester fabric is the most versatile and the strongest of fabric options; it's available as a mono or a multifilament. Monofilament polyester works well with water- and oil-based materials, making it perfect for printing slips and glazes on ceramics. For a long print run or a large frame, use multifilament polyester fabric; the weave will remain tight under such conditions.

Screen fabric comes in various mesh counts to suit the needs of your particular project. A lower mesh count has fewer and larger openings per square inch (or centimeter), allowing thicker slips or glazes to flow through it. A higher mesh count fabric has many more smaller openings in the fabric, and the result is better detail. It also lets you use thinner slips and glazes. Mesh between 120 and 260 works best, but

you'll have to do some tests, since the best size should be relative to the consistency or particle size of the slips or glazes you intend to use. For the techniques shown here, I purchased 220-mesh monofilament polyester fabric from a commercial screen-printing supplier and stretched it onto pine frames. It's a good in-between size to start with, and works well with overglazes, underglazes, and slips.

Assembling the Frame

Align the wooden boards so the mitered ends face each other; use a carpenter's square to assure they form 90-degree corners. Apply a bead of wood glue at the center of each mitered end to strengthen the joints (see photo 1). Firmly press the corners together and wipe away any excess glue. Secure the framing strap around

Christina Antemann
Handle, 2005
9 x 6 x 6 inches (22.9 x 15.2 x 15.2 cm)
Slip-cast porcelain; cone 10; decals, cone 014
Photo © Mark LaMoreaux

Materials for a screen-printing frame: mitered 1x 2-inch (5 x 10 cm) wooden boards, wood glue, a clean rag for wiping glue, a framing strap, a carpenter's square, corrugated fasteners, pliers, and a hammer

1

Dan Anderson
Water Tower Teaset, 2005
7¼ x 8½ x 4 inches (18.4 x 21.6 x 10.2 cm)
Thrown and slab-built stoneware; wood
fired, cone 10; silk-screen decals, cone 017
Photo © Jeffery Bruce

the frame (see photo 2). If more wood glue oozes out of the corners, wipe it away with a rag.

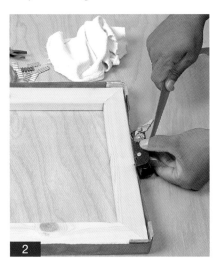

Corrugated fasteners inserted in each corner will further strengthen the frame. Use two pairs of pliers to bend all eight corrugated fasteners into L shapes (see photo 3). Place a fastener, teeth down, across the seam of each joint. Make sure they're not too close to the inner or outer edges, so as not to split the wood. Hammer each fastener straight down into the wood so its top edge is flush with the wood surface (see photo 4). Check the corners with a carpenter's square during the process to make sure the frame hasn't shifted.

Turn the frame over and hammer in the remaining four fasteners. To avoid splitting the wood, position them slightly above or below the ones on the opposite side. Leave the frame in the clamps for at least an hour, though it's best to let the glue set overnight. Check the bottle for suggested drying times.

Sealing the wood prevents warping and lengthens the life span of the frame. Any oil-based shellac or polyurethane wood sealer will work. Unclamp the frame and place it on some newspaper. Use a 2-inch (5 cm) brush to apply an even coat of the sealant on the top and sides of the frame. Let it dry, then coat the remaining surfaces. Two even coats will be enough to protect the wood from water damage and warping. Let the frame dry for 24 hours after the final coat is applied.

Janice Kluge
Chairs, 1998
10 x 10 x 1 inches (25.4 x 25.4 x 2.5 cm)
Slip-cast white body; electric fired, cone 6; gold,
lead, Lazertran decal
Photo © Lee Isaacs

Stretching the Fabric

Center the frame over the screen fabric. Trim the fabric so it's 1 inch (2.5 cm) larger than the frame's outer edges, or at least enough to wrap the frame sufficiently, as shown in photo 5.

It's better if the fabric is cut too wide rather than too narrow. Any excess can be trimmed later. Orient the weave of the fabric so that it's square to the frame pieces.

The screen must have a very taut surface for successful printing. Spray the fabric with water before stapling (see photo 6) and keep it wet until you've fully adhered it; as it dries, the fabric will wrap the frame even more tightly.

Starting in the middle of one side, shoot three staples, close together and at an angle, through the fabric and into the frame (see photo 7). The fabric is less likely to tear or rip—either now or later—when the staples are angled. Make sure they're fully flush against the fabric and wood; if necessary, use a hammer to make them so. Rotate the frame 180 degrees and, while pulling the fabric firmly taut, insert three more staples. Rotate the frame 90° at a time to tack the fabric in place.

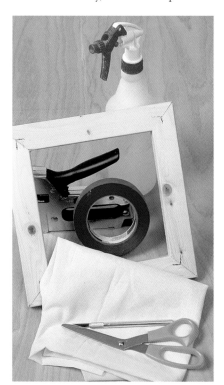

Materials for stretching the fabric include a spray bottle with water, polyester monofilament screen fabric, scissors, a craft knife, a staple gun, and waterproof tape.

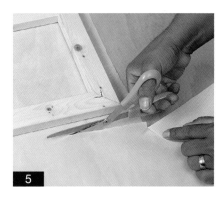

5

6

7

Finish the job by adding staples, a few at a time on each side, until you've worked your way out to the corners. Trim any excess fabric with a craft knife (see photo 8). If the fabric feels loose or it sags, restretch it. As with all new things, this process may take a little practice before getting it right.

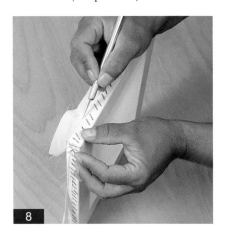

Cleanup

Once it's stretched, clean your new screen to rid the fabric of any contaminants. Many problems affecting clarity of image can be traced back to a dirty screen. Also, new, unused screen fabric may have some unseen fabric weaving–machine oil still in the mesh. Always clean your screen before putting an image on it—it's simply a good practice to follow. Cleaning can be done in a sink or outside with a garden hose, using a degreasing detergent and a soft brush.

Wet the screen with water and lightly sprinkle both sides with a degreasing detergent (either a special screen fabric degreaser or any powdered dishwasher detergent that contains trisodium phosphate). Use a nylon-bristle brush, similar in size and shape to an oversize toothbrush, to gently scrub the fabric in circular motions. Be sure to get into the corners. Rinse both sides thoroughly with water, then allow the screen to completely dry. Cleaning a screen thoroughly also extends the life of the fabric and the clarity of future images.

Matching Image to Method

Not all transfer methods are created equal, but that's a good thing because not all images are alike. Choosing the best method to apply the designs to the screen fabric depends on the complexity of details or elements in the image, as well as its tonal qualities.

Images that work best with a light-sensitive emulsion have a lot of visual elements, complex details, or multiple values. The lights and darks, or *values*, in any image—whether it's a color or black-and-white photograph or drawing—make up its *tonal scale*.

Simple images can be applied to the screen fabric with either the drawing-fluid or stencil-film method (see pages 24–26). These are limited to just a single color—usually black—on a white ground (or vice versa), with few or no in-between values. These values represent the extreme ends of the tonal scale, so the images are called *high contrast*, or *line art* (see photo). Simple art is typically bold and graphic, with fewer visual elements or details; it can even be merely a silhouette. Black type or ink on a white page is also considered line art.

Line art, whether color or black-and-white, is always boldly graphic.

Light-Sensitive Emulsion

A light-sensitive emulsion is a miraculous thing. While in its liquid state, the emulsion is applied to the screen fabric and, with sufficient exposure to light, it's transformed into a hardened film. If your original image was transferred to a clear support material (such as a sheet of inexpensive acetate), parts of it are opaque and able to block light. This *positive*, or *transparency*, is made from a drawing, a photograph, or lettering—almost any design at all—that has been sketched or photocopied (or otherwise printed) onto the acetate (see photo 9). An image for a positive can be imported into a computer using a scanner or digital camera, and further manipulated or resized with a variety of computer software programs. If you don't have a scanner or digital camera, another option is to download copyright-free photos or images from the Internet. You can arrange several images together on one sheet of acetate; just be sure to leave ½-inch (2.5 cm) borders around each one. Borders make it easy to separate the images when you're ready to apply one of them as a decal.

First, the positive is placed in tight contact with the sensitized screen and exposed to a photoflood bulb held in a reflector. (Daylight works too, but that often creates more problems than you'll really want to deal with.) The positive or black areas on the acetate block light, while the clear areas (*negative spaces*) allow it to pass through. Areas of the screen that received light become hardened emulsion in the fabric, which then can be used as a stencil in the screen-printing process.

Unexposed emulsion is washed out and the open areas create a negative version of the original image: the dark spaces are now open, empty spaces on the screen, and the light or white areas have been rendered into hardened emulsion. Colorants are then forced through those open areas with a squeegee to the surface just below the screen. In this way the image is transferred—and reversed back to its original, positive rendering.

A suitable original image must have good contrast and clean, sharp lines. Shading or subtle halftones are more difficult to work with, so when you're first learning this process start with a high-contrast original. Avoid the sort of low contrast that is created when predominant colors, such as red and green, have similar values.

A digital image, whether imported from a camera or scanned, can be inkjet-printed directly onto an

9

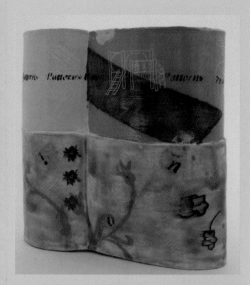

acetate sheet. Always check to make sure your acetate sheets are compatible with your printer or copier. This is particularly important because some acetates melt in the wrong machines or printers and can cause permanent damage to them. A local copy store can make the transparencies for you and will have just the right material for their copiers. Remember, positive or black areas must be fully opaque to be able to block the light. If you're unsure about their degree of opaqueness, sandwich two identical transparencies.

Coating the Screen

A screen-printing emulsion is usually made with low-sensitivity *diazo salts*. A diazo emulsion creates a sharp image, and the sensitized screen can be stored (in a darkroom-type environment,

away from light and heat) for up to several weeks before exposure.

You can make a "darkroom" out of a cabinet, a closet, a cardboard box, or any other container large enough to hold your screen. It should effectively block all light from entering it; otherwise, it will expose the screen prematurely. Use duct tape to cover any holes or cracks where light might pass into the dark environment. I've found that a sturdy cardboard box with taped-up corners—placed in a dark room or closet, just to be sure—works just fine.

Coat and expose the screen in a dimly lit room. I use a single 40-watt bulb in an otherwise dark room; there's just enough light to see what I'm doing. If the room is brightly lit, the emulsion starts to harden right away. The coating

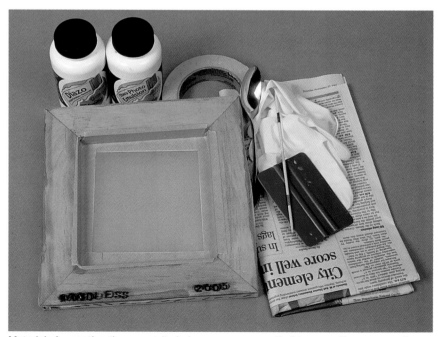

Materials for coating the screen include newspaper, a roll of tape or other spacer, diazo photo emulsion, a stirring stick, latex gloves, a spoon, and a squeegee.

process itself is very quick; just put the coated screen into your light-tight space immediately. You can also replace any light fixture's regular bulb with a safelight bulb from a photo supply store. Don't take a chance by applying the emulsion in one room and dashing, screen in hand, through a brightly lit room or hall to the darkroom. You just can't outrun light.

Use newspaper to protect your work surface from emulsion drips. With the

screen side down, prop up the end of the frame closest to you with a roll of tape or other spacer. The coated fabric can't touch any surface or it'll stick, and you'd have to rinse it out and start over again.

A diazo emulsion generally consists of two parts: the emulsion and the diazo sensitizer. Usually, cold water is added to the sensitizer before it's stirred into the emulsion. Different brands have different mixing instructions, so read and follow the instructions carefully. Only when the two parts are mixed does the emulsion become sensitized to light. Some kits also include an emulsion remover for cleaning your screen.

Wear latex gloves during the coating process to protect your hands and to make cleanup easy. Use a spoon to lay a thick bead of emulsion at the far end of the screen (see photo 10). Pull a thin, even coat of the emulsion across the screen with the squeegee (see photo 11). Use a slow and steady movement, with even pressure on the screen.

A single, even coat works fine on smaller screens (under 10 inches square [25.4 cm]), but the manufacturer may suggest an alternative coating technique for larger screens. Apply two coats on larger screens, or if the image has fine details or a wide range of tonal values. Check the manufacturer's guidelines for more information about the number of coats they recommend for the best printing detail. With multiple coats you will probably be instructed to alternate applications on both the inside and outside surfaces of

Michael T. Schmidt
Gulf Pitcher, Cup, Funnel and Base, 2004
13 x 9 x 6 inches (33 x 22.9 x 15.2 cm)
Wheel-thrown white earthenware; oxidation fired, cone 1; slip-cast porcelain, cone 6; laser print transfer, cone 06; cast struct-lite base
Photo © artist

the screen fabric, but always make sure that the final coat is on the inside. Put the excess emulsion back in its bottle and store it for future work.

Dry the coated screen, in the dark, for 12 to 24 hours. The time varies, depending on ambient humidity, air circulation, and room temperature. Check the emulsion container for estimates on drying times, as well as for appropriate storage temperatures and shelf life. Excess sensitized emulsion can be refrigerated and stored for up to eight weeks.

After the emulsion has been applied, elevate the frame with small blocks at the corners so that the screen can dry horizontally, fabric side down, without touching any surface. For satisfactory results the emulsion must dry evenly and flat in the fabric.

Exposing the Image

Burning is the process of exposing a screen coated with light-sensitive emulsion to a strong light source. A 250-watt photoflood bulb in a clamp-type fixture, directly centered over the screen, works well. Be sure the fixture has a porcelain socket so it can withstand the heat generated by the bulb. Whether the light is suspended by the cord or clamped to a small light stand, be sure the bulb's end is parallel to the screen; do not angle it.

Arrange the exposure area's setup first so you can position the sensitized screen without undue delay. At this point exposure to anything more than very low light levels will begin to harden the emulsion, so you'll want to be able to work efficiently. If possible, make the exposure in the same dimly lit area where you coated the screen. Be sure to have all the items ready so the process goes quickly and smoothly.

Place black paper or board on your worktable. It keeps the photoflood light from reflecting off the table's surface and back up through the underside of

the screen. Adjust the bulb so that it's 12 inches (30.5 cm) away from the screen fabric (for frames larger than 10 x 10 inches [25.4 x 25.4 cm], see the exposure chart on page 106).

Since the emulsion will soon begin to harden—even though the light is dim—time is of the essence in these next steps. With the photoflood light off, quickly remove the sensitized screen from its dark storage and place it, fabric side down, in full contact with the clean black work surface, in fewer than 60 seconds. Position the transparency, right side up, on the screen; the image should

Materials for making the exposure: a black sheet of paper or black illustration board, a clamp-style porcelain light socket with a 250-watt photoflood bulb in a 15-inch-diameter (38.1 cm) aluminum reflector, a light stand, your chosen image on a sheet of acetate, a sheet of glass or clear acrylic, the sensitized screen, and a timer

Far left: **Heeseung Lee**
Blue Landscape Teapot with Bird in Flight, 2004
7 x 10 x 4 inches (17.8 x 25.4 x 10.2 cm)
Slab-built terra cotta with slip and oxide transfer; electric
bisqued, cone 1; glaze fired, cone 06; multifired decal and
luster, cone 018 to 020
Photo © John Carlano

Left: **Paul Scott**
The Scott Collection, Cumbrian Blues(s) Dounreay, 2005
10 x 15 inches (25.4 x 38.1 cm)
Old Czech porcelain platter; screen printed, in-glaze decal
collage, gold rim
Photo © artist

Right: **Eric Mirabito**
Power Drill, 2005
10 x 8 x 4 inches (25.4 x 20.3 x 10.2 cm)
Wheel-thrown and altered earthenware; majolica,
electric fired, cone 017; found object, silk-screened
china paint decals
Photo © Guy Nichols

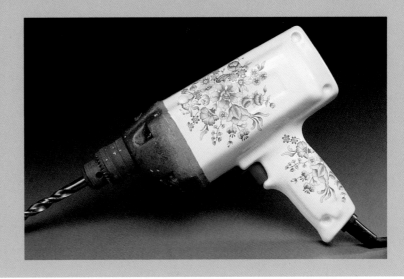

not be flopped. If it contains words or letters, they should be right-reading (that is, not reversed). Lay the glass plate or acrylic sheet on top of the transparency so that the transparency is evenly weighted and fully contacts the screen (see photo 12). You can weight down the corners of acrylic (but not so as to block any part of the image or create shadows), but glass works better because it's heavier. Any space between the transparency and the screen will cause the image to be blurry.

Turn on the light and expose the screen for 12 minutes (see photo 13); use a timer. Remove both the glass and the transparency from the screen. Exposure time can vary from 1 to 3 minutes, depending on the thickness of emulsion coating, the type and age of the photoflood, the size of the screen, and the distance between the light source and the screen. You can make test strips on a piece of fabric mesh to better identify your exposure time.

Immediately take the screen to a rinsing area (not outdoors! Any unexposed areas of the screen would harden before you could get the emulsion rinsed out). Direct a forceful spray of cold or tepid water through both sides of the screen, until all of the unhardened emulsion areas have washed away. It may take a few minutes. You now have a more or less permanent image fixed into the screen fabric itself; the screen has become a stencil. Let the screen dry thoroughly in a horizontal position.

Check for pinholes in the hardened areas after the screen is dry. These can be covered with masking tape during the printing process. The image is ready to be transferred to a decal or a clay or plaster slab using slips or glazes (see photo 14).

Unless you want the image to remain permanently in the screen, remove hardened emulsion when you're done. You can use the screen cleaner and degreaser that's usually included in a diazo emulsion kit. The cleaner is brushed on both sides of the screen to break down the emulsion. Use warm water and a stiff nylon brush to scrub the screen clean.

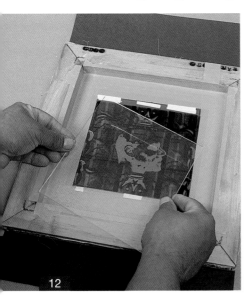

12

13

14

Left: **Jessica Broad**
Book Series #4, Cracked Layers, 2003
2 x 7 x 5 inches (5.1 x 17.8 x 12.7 cm)
Slab-built porcelain; electric fired, cone 6;
laser-printed decals from digital photographs and
scanned drawings, unglazed, cone 6
Photo © artist

Right: **Cherie Westmoreland**
Family Portraits, 2004
6⅜ x 6¼ x 2¼ inches (16.2 x 15.9 x 5.7 cm)
Slab-built stoneware; electric fired, cone 7; laser
print transfer
Photo © artist

Drawing Fluid

For artists who enjoy drawing, a method that uses drawing fluid and screen filler is a simple, direct way to put a unique and personal image on a screen. A sketch is made on the screen with drawing fluid. Screen filler is then applied, though the drawing fluid resists it. The filler hardens, in a way similar to the photo-sensitive method just demonstrated. The principle is a lot like using wax resist to mask off areas on a pot.

Spread newspaper to protect your work surface and place the screen, with the inside facing up, atop some spacers. The drawing will print exactly as you execute it on the fabric. If it doesn't matter that the image will be reversed when printed, then you can work on the outside—in which case you won't need any spacers since the frame itself keeps the fluids from touching the newspaper. In this demonstration I wanted the writing to be legible in the final print, so I worked from the back side, or outside, of the screen.

Drawing the Image

Thoroughly mix the drawing fluid to an even consistency so you'll get a smooth application. Select a brush that's appropriate for the type of line you want and paint the fluid right onto the screen image (see photo 15). If you wish, make a light pencil drawing on the screen first, then paint over it with the drawing fluid. Apply a single, even coat of the fluid for best results. If applied too thickly it'll take longer to dry and will be harder to wash out. Dry the screen in a level, horizontal position; you can use a fan to quicken the drying time.

Stir the screen filler, then pour a bead, on the same side that you applied the drawing fluid, at one end of the screen. Squeegee a smooth, even coat of the filler across the screen (see photo 16). Multiple coats aren't recommended because the screen filler can dissolve the drawing fluid. Let the screen dry horizontally.

Washing Out

Spray warm water through both sides of the screen, concentrating on areas where the drawing fluid was applied. The drawing fluid washes out, creating your drawing as a negative image (see photo 17). Screen filler will be removed later with a cleaning detergent and a stiff brush. Carefully read and follow cleaning instructions on the container of the brand of screen filler you use.

15

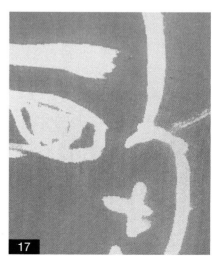

16

17

Stencil Film

Stencil film works best with images that have just a few details. Line art—or any image with mostly open spaces—works well for this screening method. Stencil film is a vinyl material with an adhesive backing that you attach directly to the screen fabric. It's not a permanent adhesion, but it will last for about 30 or so prints.

18

A stencil reverses the image, so any text or numbers must be reversed on the screen in order to be legible on the clay.

Cutting the Stencil

Measure the frame and use a craft knife to cut the film to a size slightly larger than the screen area (see photo 18). Secure the film with masking tape, backing side down, to a flat surface. Draw a border on the film with a pencil and ruler to outline the size of your screen's printing area. With a sheet of carbon paper under it, tape a paper image to the film. Transfer the image to the film by tracing it with a pencil (see photo 19).

Follow the transferred lines with just the tip of the craft knife, pressing hard enough to cut the vinyl, but not so

hard you cut through the backing paper. Use the blade's tip to lift one edge, and slowly peel away the image section (see photo 20). Notice that there are some free-floating areas in my image (the open spaces in the chair back); I took extra care to make sure my cuts at all the corners were complete so that I didn't inadvertently pull away those open areas too.

Securing the Film

Once you've removed all of the stencil film from the image, peel the backing paper from a sheet of clear vinyl. Apply it, adhesive side down, over the entire image area. The vinyl sheet keeps the image's separate parts (if it has any) in place (see photo 21). Gently rub the surface to assure good adhesion. Flip over the stencil and slowly remove the stencil's white paper

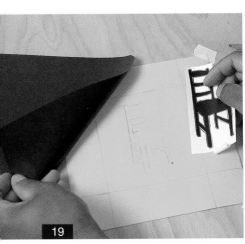

19

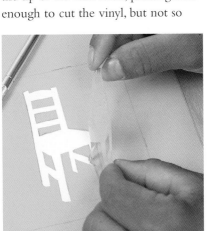

20

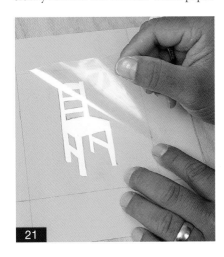

21

backing to reveal the adhesive side of the stencil film (see photo 22).

Place the screen, fabric side down, onto the film (see photo 23). Cut a small square from the stencil's backing paper and use the smooth side to rub the screen until it adheres well to the film under it (see photo 24). Turn over the screen and slowly peel away the clear vinyl (see photo 25). Seal the edges of the stencil (on both sides of the screen) with painter's tape (a low-tack, waterproof masking tape), as shown in photo 26. The waterproof tape helps keep ceramic underglazes from getting under the film during printing and cleaning.

You can simply peel the stencil film off the screen when you're done printing the image, and clean the screen with a scrubbing detergent and a stiff brush before using it again.

22

Top: **Brendan Tang**
Out in the Cold, 2005
11 x 12½ x 5 inches (27.9 x 31.8 x 12.7 cm)
Wheel-thrown and altered stoneware, cone 6;
Lazertran, unfired; commercial decals, cone 018
Photo © artist

Bottom: **Amy M. Santoferraro**
Road Show Prop #SRJ93572, 2004
4 x 6 x 3 inches (10.2 x 15.2 x 7.6 cm)
Slip-cast porcelain; electric fired, cone 6; decal
and gold luster, cone 018
Photo © artist

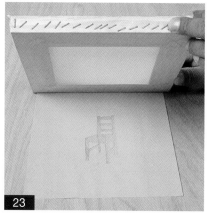
23

24

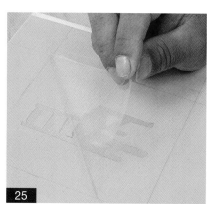
25

26

Decals

My introduction to the decal process was simple, exciting, and probably familiar to most folks. When I was young, I remember getting tattoo decals out of gumball machines and candy boxes. It was great fun to use this quick process to transfer images onto my arms, legs—even my forehead. Little did I know that, years later, I'd be using a similar technique on clay. There are three different kinds of decal methods and materials. The one you use depends on the type of image you want to transfer.

Oil-Base Overglaze

Decal paper is a commercial product coated on one side with a thin, clear layer of gelatin. Overglaze is screened onto this gelatinous layer, allowed to dry, then sealed with a cover coat of lacquer (see figure 1). The overglaze can be applied to decal paper by screen printing, hand painting, or a combination of the two.

Cover coat
Overglaze layer
Gelatinous layer
Backing paper

Figure 1

The ceramic "ink" used for screening the decal is oil-based china paint, or *overglaze*. Oxide or stain mixed with flux and a binder are the components of a typical overglaze. These decals are applied onto a glazed, fired ceramic surface and refired within a range of cone 023 to cone 018 (whatever's right for the overglaze)—just hot enough to burn away the clear layers and permanently melt the overglaze onto the glazed surface.

You can fire several decals with different firing temperatures on the same piece. Be sure to plan the design so as to be able to fire the overglazes that require the hottest temperatures first, so that the cooler-firing ones won't burn away. Some overglazes can't be refired at all (typically warmer colors, such as reds and oranges), so be sure to take that into account, too.

Oil-base overglazes are available in a wide variety of rich and vibrant colors, but some brands contain lead. Tube overglaze is ready to print; the dry powder form must be hand-mixed with an oil-base printing medium. Mixing your own is simple, and both varieties have the same richness of color and are easy to screen.

Rain Harris
Rapunzel, 2004
13 x 13 x 21 inches (33 x 33 x 53.3 cm)
Slip-cast and altered cone 6 porcelain; decals, multifired, cone 017
Photo © John Carlano

Top: **Victor Spinski**
Ashtray on Slab, 1999
9 x 9 x 3 inches (22.9 x 22.9 x 7.6 cm)
Slip-cast and slab-built low-fire white clay;
electric bisque fired, cone 04; glaze, cone 06;
decals, cone 018
Photo © artist

Bottom: **Bill Brown**
Twenty-Four, 2004
24 plates, 12½ x 15 inches
(31.8 x 38.1 cm) each
Press-molded stoneware; electric fired, cone 7
biscuit, cone 04 glaze; silk-screen enamel and
digitally printed decals, cone 015
Photo © artist

Single-Color Process

Single-color printing is the most basic method for making an overglaze decal. Here, artist Victor Spinski shows how to create a two-color image with the single-color printing method. On

To mix ceramic ink from powdered pigment, you'll need newspaper, a sheet of glass, latex gloves, powdered overglaze pigments, a stiff metal spreader or wide palette knife, oil-base decal medium, a glass jar, and a glass container.

Safety

There are lead-free overglazes available, but most overglazes use lead as a flux. Take the time to read the labels of your materials closely for product warnings and recommendations. Never use materials containing lead on any ceramic piece that might ever come into contact with food. In fact, any use of lead-base overglaze on dinnerware is illegal in the U.S.

Oil-base overglazes also require that you use strong solvents: mineral spirits for general cleaning and lacquer thinner for cover coat cleanup. All the products used in this book were chosen because they work well and under the right conditions are safe to use. Follow these simple guidelines:

• Wear latex or rubber gloves whenever you're mixing or cleaning up emulsions, detergents, degreasers, oils, or solvents. Protect your skin and eyes from direct contact with them.

• To avoid vapor buildup, always ensure that your workspace has good ventilation. Wear a mask or respirator specifically designed to protect your lungs from solvent vapors, and follow the directions, and safety and disposal instructions for all products.

The same basic setup for mixing overglazes is used to do the printing, too. In addition, you'll need decal paper, scissors, 1-inch-wide (2.5 cm) masking tape, the screen with image, a spoon, the overglaze, a squeegee, an acrylic sheet, and cardboard or illustration board.

separate sheets of decal paper, he prints the same screen in two different colors, then cuts them apart and reassembles the parts to make one multicolor image. This technique can be used for any image that has simple shapes that don't overlap. It's an easy way to make a complex image.

Mixing Dry Pigment

Screening an image onto decal paper requires only careful preparation to assure a smooth printing session, or *run*. Choose a roomy work area with good ventilation and have everything you need close by. Cut the decal paper to a size appropriate for the screen. Have a few sheets of cardboard or some other vertical surface close by, where decals can be taped up to dry after being printed.

For screen-printing purposes, oil-base overglaze should be mixed to the consistency of honey or heavy syrup. Make just enough ceramic ink for

your project because the shelf life after it's mixed is only about seven days. Keep a few extra sheets of decal paper handy to print any excess ink you may have, or save the ink in a glass jar and use it for another printing run later in the week.

Lay a sheet of glass onto spread-out newspaper or kraft paper. Pour a small mound of the dry pigment onto the glass (see photo 27). Use a metal

27

Susan Feller Mollet
Past and Present, 2003
16 x 11 x 8 inches (40.6 x 27.9 x 20.3 cm)
Coil-built black stoneware; cone 10 reduction; multifired glazes, cone 04; photocopy decal, cone 10
Photo © Harrison Evans

29

Dalia Lauckaité-Jakimavičiené
A Feast, 2003
21½ x 25½ x 1 inches (55 x 65 x 2.5 cm)
Slab-built earthenware; electric fired, cone
01; commercial decals, china paint, gold,
luster, cone 015
Photo © Vidmantas Ilčiukas

spreader to clear an opening in the center (see photo 28). Slowly pour decal medium into that space, then

mix the two materials together (see photo 29), scooping the spreader under an outside edge of the pigment

and folding it over, into the middle. Continue until the two materials are evenly mixed and the color is uniformly opaque. You may need to add more pigment for the proper consistency and opaqueness (see photo 30). Victor mixed both of his glazes on the same sheet of glass and stored them in glass jars with lids (see photo 31).

Clean the glass and spreader with odorless mineral spirits and a cloth rag (see photo 32). To reduce fumes in the work area, do the cleanup immediately. Once the overglaze is mixed, the screen printing must proceed quickly, with only short breaks between prints.

Cut the decal paper to size, making sure you have at least three or four extra sheets so you can proof for quality.

28

29

30

31

Paul Scott
Scott's Bumbrian Blue(s) Vignette, 2005
13 x 12 x 10 inches (33 x 30.5 x 25.4 cm)
Earthenware ceramic forms; tin glazed, in-glaze
decals; decals on porcelain tiles
Photo © artist

Secure a sheet on the tabletop with masking tape at the corners (see photo 33). Align and center the screen over the decal paper. Prop up the near end of the screen with a roll of masking tape. The fabric should contact the decal paper only during the actual printing. Using a plastic spoon, distribute a thick bead of overglaze, slightly wider than your image, across the top of the screen (see photo 34). The bead should stand up and appear to be firm (see photo 35); if it spreads out, it's too thin.

Lower the screen onto the decal paper. Hold the squeegee at a 45° angle, pressing firmly as you pull the overglaze across the screen. Take care not to let the screen shift its position once it has made contact with the

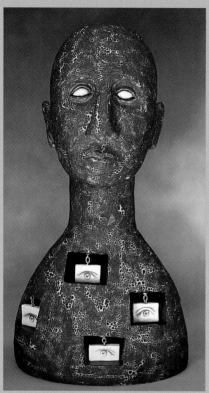

Kina Crow
Focus, 2005
20 x 10 x 8 inches (50.8 x 25.4 x 20.3 cm)
Coil-built stoneware and porcelain;
electric fired, cone 6; photocopy transfer,
crawling glaze
Photo © artist

decal. Reverse the direction of your pull and squeegee the overglaze back across the image (see photo 36). The color should be even and cover the entire opening in the screen (see photo 37).

Lift the frame and move it to one side, with one end propped up to keep the inked fabric off the work surface. Tape the decal to a rigid horizontal support material, such as an acrylic sheet, cardboard, or illustration board. Overglaze dries from the outside down to the paper, so it may look dry after 10 minutes but still be wet underneath. You can use a fan to quicken the drying time. The glaze may run if it dries in

Screen Print Troubleshooting Guide

Most problems are easy to notice and simple to correct. Here are solutions for some of the most common problems. Print as many test decals as it takes; sometimes three or four prints are needed before you get it right.

- **Uneven color distribution.** Use more even pressure on the squeegee (see photo, top right).

- **Splotchy color.** A too-thick overglaze can clog the screen with excess ink and make the image look blurry at the edges (see photo, bottom right). Add a few drops of commercial thinning product or paint thinner and stir well. The streak in the S was caused by a clump of undermixed dry pigment. Here's a case where more than one problem—thick glaze and undermixing—occurred in the same print.

- **Thin, runny overglaze.** Add decal medium so that the overglaze isn't so thin that it runs into areas where it isn't wanted (see photo, bottom left).

- **Weak or pale color.** Add more pigment, and mix it well.

- **Tiny dots of unwanted color in blocked areas.** You may have some pinholes in the blocked-out areas. Quickly—because the glaze is starting to dry—apply a piece of tape over the problem. Taping up all of the non-printing (blocked, or closed) areas of the screen before you start printing protects the fabric and makes cleanup easier too. Use either regular masking tape or painter's tape.

- **Smudged or blurred images.** The decal paper shifted during your pass with the squeegee, or the screen shifted slightly as you positioned or lifted it.

a vertical position—especially if the image has large areas of solid color—so keep the decal horizontal for at least 30 minutes. Text and line art can be dried vertically if the application of overglaze isn't too heavy.

Examine this first image for any printing corrections that may be needed (see the troubleshooting guide at left). Once you've finished the print run for one color, immediately clean your screen. Overglaze soon dries in the fabric, clogging the open, unblocked parts of the

image. Let the screen dry for an hour before printing a new color. Allow the decals to dry overnight before applying the cover coat.

Applying the Cover Coat

A cover coat is a layer of lacquer, the final finish of a decal "sandwich." The cover coat should completely cover the image, and is applied only when the overglaze is thoroughly dry.

Assemble the materials shown in photo 38 and tape the decal to a newspaper-covered work surface.

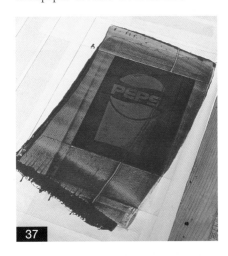

36

37

38

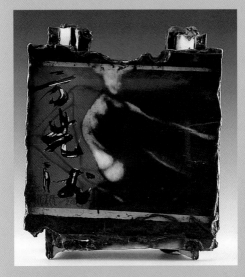

Top: **Tracey Scheich**
Reach, 2004
8 x 13 x 3 inches (20.3 x 33 x 7.6 cm)
Slab-built porcelain; electric fired, cone 6; digital image on Lazertran
Photo © artist

Bottom: **Bill Brown**
Play the Scottish Card, 2000
18½ x 14 x 2 inches (47 x 35.6 x 5.1 cm)
Slip-cast white earthenware; electric biscuit fired, cone 4; glaze, cone 04; silk-screen enamel decals, cone 015
Photo © artist

33

Scott Bennett
Saucer, 2005
18 x 5 inches (45.7 x 12.7 cm)
Thrown earthenware; electric fired,
cone 05; decal and luster, cone 017
Photo © artist
Collection of Red Dot Gallery

Position the open screen over the decal and apply overlapping pieces of tape to the entire area around the image, plus a ¼-inch (.6 cm) border (see photo 39).

Cover coat is thinner than overglaze and has to be spread a little quicker, in a single smooth stroke. The screen must be lifted off right away, or the colored ink, softened by the cover coat, could be absorbed into the screen, leaving a shadow of the image on the printing side. That ghost image could then be transferred to the next decal when it's cover-coated.

Screening the cover coat is the same process as screening the overglaze. Pour a bead across the screen opening. Using a spoon (see photo 40), pull the cover coat across the image (see photo 41), and remove the screen right away (see photo 42). Tape the coated decal onto a horizontal drying board. Let cover-coated decals dry for 24 hours before applying them to a surface to be fired.

One even coat is all that's needed. If the cover coat is too thick, the decal might be too stiff to adhere well to the clay; too thin and it might tear during the soaking and application stages. This final stage of making the decal is very important.

Multicolor Images from Single-Color Decals

A simple way to make a complex image is to assemble single-color decals into a new multicolor image (see photo 43), like a decal collage on the ceramic surface. Victor printed the same image in blue and red, then cut and reassembled them to make a two-color image on a glazed commercial tile.

An overglaze decal must be applied to an already-fired, glazed surface. The temperature of the glazed surface can

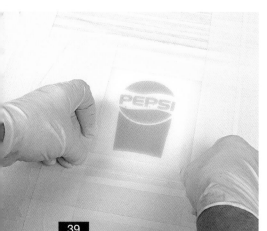

39

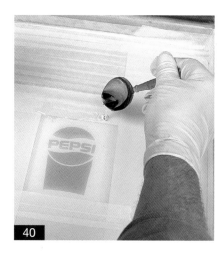

40

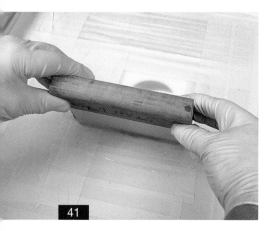

41

42

Amy M. Santoferraro
Quiet Is the Rabbit on the Hill, 2005
4 x 5 x 3 inches (10.2 x 12.7 x 7.6 cm)
Rabbit: slip-cast porcelain; reduction fired, cone 10;
brick: solid terra cotta; electric fired, cone 04; photocopy transfer, decal, cone 018; pillow: slip-cast
porcelain, electric fired, cone 6
Photo © artist

range from cone 06 to cone 10. Clear or white glaze is most commonly used, but you can use almost any color you find suitable. Smooth, glossy finishes work the best, but matte or textured ones can work, too—with a little testing.

Cut out the individual parts of the image, leaving at least a ¼-inch (.6 cm) border around the glaze. Align the pieces on the surface (see photo 44). Submerge the decals in a container of cool water and let them soak for 30 to

45 seconds. A water-soaked decal curls first, then flattens a little. Once the image has loosened itself from the backing paper, remove it from the water, and position it face up on the glazed ceramic surface. Slide the backing paper out from behind the decal itself (see photo 45). Use your fingers to carefully smooth out any wrinkles and to make any last adjustments to the image. Dab away excess water with a soft cloth (see photo 46). Tissues or cotton balls can be used, but be careful not to leave any lint on the decal.

43

44

45

46

Lenny Dowhie
Australian Impression, 1996
20 x 8 x 5 inches (50.8 x 20.3 x 12.7 cm)
Cast and constructed whiteware/commercial Australian slip; multi-fired electric in oxidation, cones 04, 06, 17; commercial glaze, decal collage
Photo © Lenny Dowhie

Multicolor Process

To create a multicolor decal, you have the option of making separate screens for each color layer, or you can streamline the process by putting all the colors on different parts of one large screen. To print the different colors, the frame is simply moved around on the receiving surface. If your multicolor image is smaller than 2 inches (5 cm) in any direction, I recommend that you put all the color layers onto one large screen. The one-screen approach was used for this demonstration, plus I used commercial oil-base overglazes in tubes this time, instead of the mixed-from-scratch kind.

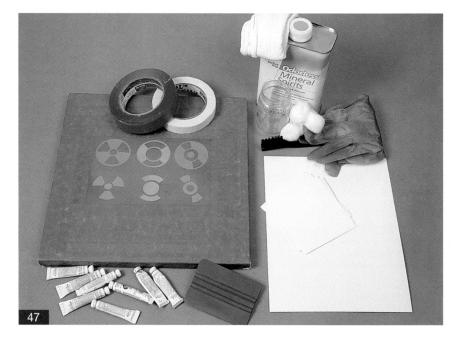

47

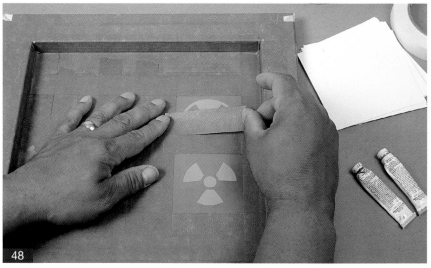

48

Sight-Registration Screening

The screen shown in photo 47 contains three two-color images, with the two parts of each image aligned in their own columns. If you carefully plan the positions of the image's separate parts (each of which represents a different color), you'll be able to easily and accurately reposition, or *register*, the screen for subsequent colors.

Mask off all the areas of the screen except the image you're going to print first (see photo 48). Align the decal and the frame by first marking the outermost position of one corner of the decal paper with tape. Small pieces of tape can secure the decal's other corners. Align the screen over the decal and mark the position of the frame's bottom right corner with masking tape (see photo 49).

Place the screen over the decal paper and extrude a bead of the first overglaze color just above the image (see photo 50). Pull the overglaze across the image to print the decal (see photo 51). Print as many additional sheets of this first color as you feel you'll need.

David Ray
The Clowns Want Your Goat for Processing, 2005
25 x 13 x 11 inches (63.5 x 33 x 27.9 cm)
Hand-built earthenware with manipulated sprig mold detail; collaged commercial and handmade silk-screened decals, hand-painted transfers, enamels, lusters, multifired, cone 03 to cone 017
Photo © Shannon McGrath

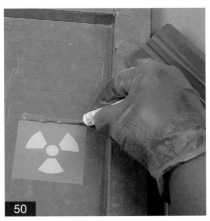

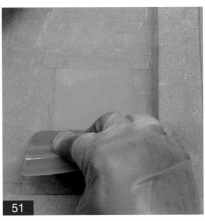

37

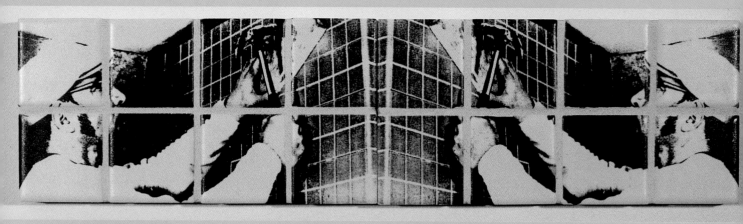

Shawn Williams
Leveling, Leveling, 2004
4 x 16 x ½ inch (10.2 x 40.6 x 1.3 cm)
Ceramic; electric fired, cone 06; silk-screened decal; grout, wood
Photo © artist

Clean the screen with mineral spirits and a soft cloth or cotton balls. A nylon brush can also be used for the more stubborn areas. Let the decals dry completely before printing the next color.

Mask all but the area of the screen that has the second color. Put a sheet of printed decal paper in place on the worktable. Hold the screen above the printed decal paper; the screen is

registered when the already-printed part of the decal image covers the screen's blocked areas (see photo 52; photo 53 shows the screen slightly out of register with the printed first color). The first decal can be used as a test print to see if the alignment is correct (see photo 54). If the second color image is nicely registered with the first one, print the rest of the decals. Tape them to a flat surface to dry; smaller ones can dry vertically (see photo 55). Let the overglaze dry for 24 hours before applying the cover coat to finish the process.

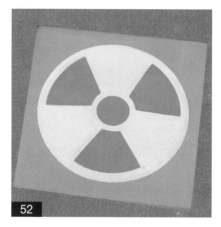
52

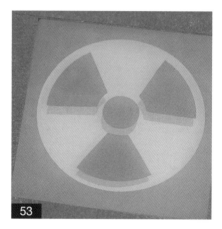
53

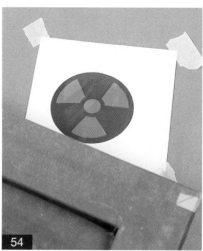
54

55

Hand Painting

If you enjoy painting and drawing, use that passion to work directly on decal paper. While it's possible to paint overglaze right onto a ceramic piece, it's not always practical for curved or vertical surfaces. A decal is easier to paint on—there's no danger of the color running or dripping. Hand painting a decal is a pretty straightforward process. You can draw a one-of-a-kind design freehand or screen print part of an image on the decal first, then fill in open areas with hand-applied color.

Paint thin coats of overglaze directly onto the decal with a soft brush made of camel hair or sable (see photo 56). The overglaze should spread easily and can be thinned if necessary. Always

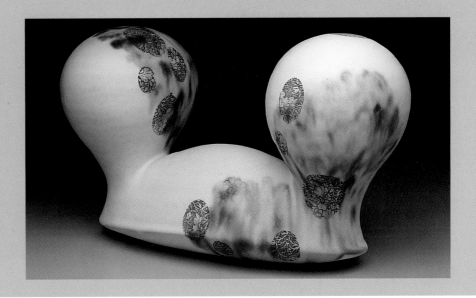

Erin B. Furimsky
Untitled, 2004
7 x 12 x 7 inches (17.8 x 30.5 x 17.8 cm)
Hand-built white stoneware; electric fired,
cone 6; computer-printed decals, cone 04
Photo © Tyler Lotz

check the overglaze label first for directions about which is the proper thinner to use, but mineral spirits typically work well. Overglazes specifically for brush painting are also available in small bottles, tubes, or trays. Use a soft cloth, cotton ball, or cotton-tipped stick dipped in thinner to remove unwanted marks.

Electric Kiln Firing

Allow at least 2 inches' (5 cm) space around the pieces in the kiln so that fumes can escape. Ventilation is very important when firing overglaze colors; fumes can affect how decals adhere and the final look of the color, especially any that contain lead. If you typically bisque- and glaze-fire in the

same kiln, first fire it empty to cone 01 to burn out any sulfur buildup in the soft brick. With an unventilated kiln, or if you're unsure of the kiln's venting capability, fire it with the lid propped slightly and have the spy holes open. Check the manufacturer's instruction book for their recommendations on how to handle fumes.

The kiln should be vented or the room itself well ventilated. Some fuming will occur in the early stages of the firing. Set the kiln on low, with the lid propped open and the spy plugs out, for 45 minutes. Turn the kiln up to medium for another 45 to 60 minutes; by this time the fumes will have dissipated. Turn the kiln to high and close

the lid; plug the spy holes when the kiln reaches 1100°F (593.3°C). When the firing is finished, let the kiln cool to room temperature (with the spy holes still plugged).

Some colors require different firing schedules than the one described here, and may also need a soaking period at the end of the firing. Keep in mind that not all overglazes fire to the same cone. Large, thick pieces, such as sculptural works are usually fired slower. Determine the right firing schedule for your work by test-firing some decals on practice pieces.

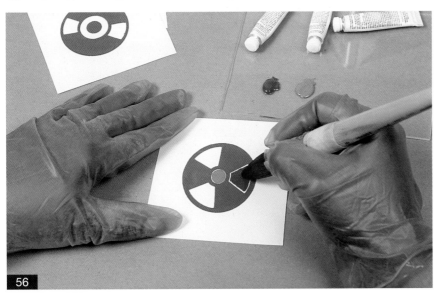

56

Laser Toner Decals

Toner creates black- to sepia-toned images from the iron-rich oxide pigment found in laser toner printers and laser toner copier machines. Images with bold lines and good contrast (including text, numbers, and characters) work best for this process. Different brands of printers and copiers have varying concentrations of iron oxide, which can affect the lightness or darkness of the image (see photo 56). Some may not even work at all. You must test-fire your first few decals. Bubble jet, inkjet, or any other types of printers will not work; they use little or no iron oxides in the inks. Use laser toner machines only.

The same decal paper used for overglazes (the kind that require a cover coat) can be used for the toner process. There's also a specialty product available called *laser toner decal paper* that doesn't need a cover coat. The toner becomes fused to the surface of this special decal paper and is ready to use immediately after printing. Both types of decal paper can be fired onto a glossy or semi-glossy fired and glazed surface. The iron-oxide image is absorbed into the glaze's surface during the firing.

The best results may be had with cone 06 to cone 02 glazes, but with a little testing, the higher temperature glazes can be successful too. Toner decals can also work on matte glazes and unglazed clay bodies if they are fired to higher temperatures—cone 02 to cone 7, depending on the clay body used and image color desired—so that the iron oxide can flux. Since the image is usually brown to reddish-brown, apply the laser decal to light-colored clay or slip. But be sure the decal firing is the last stage of the process; glazing on top of a laser decal image will dissolve it.

Making the Print

The first step is to photocopy or print an image onto a sheet of decal paper (you may need to cut it to fit the standard size required by the printer). Make clean cuts because jagged edges or angled sides won't feed through the printer properly and may cause paper jams. Load the decal paper into the paper tray or feeder of a laser toner printer, making sure to properly orient the printing side to the emulsion, which is shinier and smoother than the backing-paper side. Most copy shops can print a computer file directly to a laser printer.

A selection of toner decals

Left: **Paul Andrew Wandless**
Three Parts Water, 2005
5 x 15 x 3½ inches (12.7 x 38.1 x 8.9 cm)
Slip-cast low-fire clay; laser toner decals,
embossed linocut print; wood
Photo © artist

Right: **Richard Burkett**
Feast and Famine Platter, 2005
1¼ x 14 x 6½ inches (3.2 x 35.6 x 16.5 cm)
Slip-cast porcelain; gas fired in reduction,
cone 10; laser printer decals, cone 05
Photo © artist

In a well-ventilated area, spray a cover coat of clear lacquer over the toner image. You don't need to use liquid lacquer for this type of decal because the printing process has already partially fused the iron oxide to the paper; a layer of toner isn't as thick as screened overglaze. Typically two or three light coats will be enough. Avoid heavy spraying so the lacquer doesn't drip. Allow the decal to dry completely before applying it. Check the directions on the spray can for recommended drying times.

Applying the Decal

A toner decal is typically fired to the same temperature that the glaze underneath it was—or close to it, depending on the glaze. The glaze only needs to be soft enough to allow the iron oxide to be absorbed into the top layer of the glaze and become a permanent part of its surface. Smooth glossy finishes work the

best, but matte, textured, and bisque surfaces can work with a little trial and error, as long as they're fired hot enough to fuse. Toner decals can also be layered over each other, firing them multiple times.

Remove the decal from the water (notice how it curls and slips free in photos 57 and 58); let the excess water drip off, then align the decal on the fired and glazed surface. Slide the backing paper out from behind the decal and adjust its position as necessary (see photo 59). Use a cotton ball or soft cloth to remove excess water and to smooth out any wrinkles. For best results, the entire decal must be in direct contact with the surface.

Firing

Fire the decal in a ventilated electric kiln or in a ventilated room, keeping the spy holes open and the lid propped. Vent the fumes that are

created as the lacquer burns away. Fume buildup inside the kiln might affect the glazes. By the time the kiln reaches red heat, the lacquer is gone and you can plug up the spy holes for the rest of the firing.

The firing temperature for a toner decal can vary. A good rule of thumb is to fire the decal to a cooler temperature than the piece's glaze. If, for example, the glaze was fired at cone 6 or lower, fire the decal 4 to 6 cones cooler (the full range of cones is given on pages 108–109). Toner decals fired on cone-6 to cone-10 glazes can be fired between cone 06 to cone 04. These are only recommended temperature ranges; adjust them as needed for your particular project. The glaze surface, whether glossy, semigloss, or

57

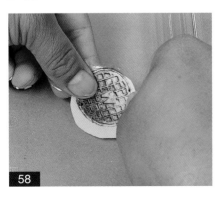

58

59

Above: **Trine Hovden**
Grandad at Hovden/Site-Specific Installation, 2002
60 x 100 inches (152.4 x 254 cm)
Industrial tiles; reglazed, electric fired, cone 05a; silk-screened, electric fired, cone 05a
Photo © Jon Nesse

Right: **Christine Julin Shanks**
Metamorphosis, 2004
15 x ⅛ x 11 inches (38.1 x 0.3 x 27.9 cm)
Slab-built porcelain; electric fired, cone 04; glazed, cone 06; Lazertran, semigloss UV protective spray coat
Photo © artist

matte, also affects the firing range too. Always test-fire.

Nonfired Full-Color Decals

A commercial decal product called Lazertran lets you create a full-color decal without having to fire it at all. There are two kinds of nonfiring Lazertran decal papers that are suitable for ceramics. One is printed in a color copier and the other in an inkjet printer. Neither type of paper should be used in any other kind of printer because the lacquer coating on the paper will melt and stick to a toner-type copier's heated drum.

The Lazertran decal image is adhered to a fired glazed surface with a solvent, but the image isn't as permanent as a fired decal; it will be vulnerable to scratches from abrasive materials. Use an acrylic sealer to help protect it. Still, this product is a fast and easy alternative for full-color images. Use it only for aesthetic or decorative purposes—it's not food-safe.

Making the Decal

Transfer your image by printing it onto a Lazertran sheet with a color copier or an inkjet printer. Print on the smooth, shiny side, and use the manual feed tray to avoid a printer jam. Wait 30 minutes for the ink to completely set before using the decal; otherwise, the ink will release in the water.

Artist Tracey Scheich demonstrates and shares some tips on how to apply a Lazertan decal to a fired and glazed surface. Tracy made her decal from her own conté-crayon and watercolor

drawing. She imported the image into her computer and used image manipulation software to select and duplicate one of its details. She printed the final image and had a copy shop put it onto a sheet of Lazertran with a color copier. Here she applies the decal to the face of her hand-built box. Its clear glaze had been previously fired to cone 04.

Applying the Decal

Trim the image with scissors, leaving a small border on all sides (see photo 60). To help with positioning, apply a little water to the surface where the

60

61

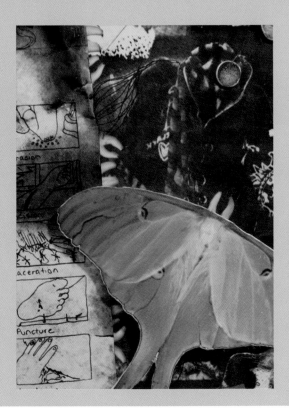

Above: **Janice Kluge, with Doug Baulos**
One Hundred Hands, 2005
8 x 3½ x ½ inches (20.3 x 8.9 x 1.3 cm) each
Slip-cast low-fire whiteware; electric fired, cone 06; glazed, painted,
ceramic decals, Lazertran, rubber stamps
Photo © Lee Isaacs

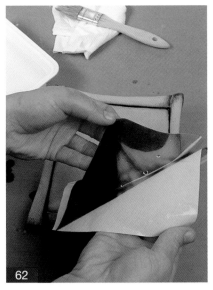

decal will go (see photo 61). Immerse and soak the decal in room temperature water until the image can slide around on the backing paper; peel it off (see photo 62). Position the decal on the glazed surface (see photo 63). Use a soft cloth to wipe away excess water and smooth out any air bubbles. Let it dry for about 15 minutes.

Adhere the decal with turpentine. (Use a workspace with good ventilation and wear latex gloves.) Use a soft brush to paint a light, even coat of turpentine onto the entire surface of the decal (see photo 64). Don't apply too much

or the image will break up, or even disappear. The turpentine softens the decal's gum substrate so that the image can be drawn into the glazed surface a little bit. Although the image bites into the surface for a very secure adhesion, it's not physically a part of the glaze itself. Lightly dab any excess turpentine away with a soft cloth. Let the piece dry for several hours. Unprinted areas of a Lazertran decal made with a color copier are transparent, but remain white on the inkjet type. To change the white to clear, apply a coat of polyurethane.

62

63

64

Christina Antemann
Place, 2005
10 x 12 x 8 inches (25.4 x 30.5 x 20.3 cm)
Slip-cast porcelain, cone 10; decal, cone 014
Photo © Mark LaMoreaux

Elissa Armstrong
The Deer, 2003
15 x 10 x 7 inches (38.1 x 25.4 x 17.8 cm)
Hand-built earthenware; electric fired, cone 04; decals, cone 018
Photo © artist

Nan Smith
A Reliquary, 2003
15¾ x 16⅛ x 21¹⁄₁₆ inches (40 x 41 x 53.5 cm)
Press-molded, modeled, and slab-built earthenware; electric fired, cone 04;
digital photo decal, laser printed, cone 09
Photo © Allen Cheuvront

Mel Robson
re:collect 2, 2004
4¾ x 23¾ x 3¼ inches (12 x 60 x 8 cm)
Slip-cast porcelain; electric fired, cone 10; custom decals
Photo © Rod Buccholz

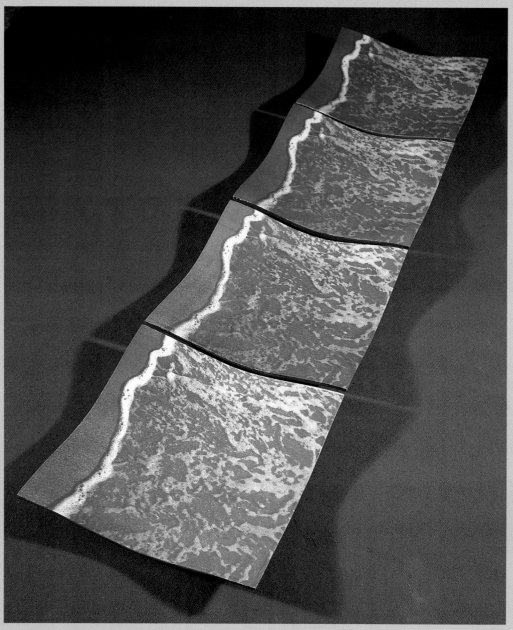

Top: **Susan Feller Mollet**
Journal Stones: Thirty-Five Days, 2003
8 x 24 x 12 inches (20.3 x 61 x 30.5 cm)
Hand-built stoneware; cone 10 reduction; silk-screened under-glaze, oxide wash
Photo © Harrison Evans

Left: **Bill Brown**
Shoreline, 2000
14 x 14 inches (35.6 x 35.6 cm)
Slip-cast white earthenware; electric biscuit fired, cone 4; glaze, cone 04; silk-screen enamel decal, cone 015
Photo © artist

Top right: **Cherie Westmoreland**
Grouping of laser image transfers, 2005
Largest, 4 x 5 inches (10.2 x 12.7 cm)
Slab-built stoneware; electric fired, cone 6; laser image trans-fer; low-fire clear glaze; iron oxide, glazes
Photo © artist

Bottom right: **Trine Hovden**
Fix Points, 2005
15 x 80 x 80 inches (38.1 x 203.2 x 203.2 cm)
Slab-built terra cotta; electric fired; silk screened onto wet clay, cone 04a
Photo © artist

Top: **Shawn Williams**
Laying tiles, 2004
16 x 16 x ½ inches (40.6 x 40.6 x 1.3 cm)
Ceramic tiles; electric fired, cone 06; silk-screened
decal; grout, wood
Photo © John Carlano

Left: **Jeff Irwin**
Call, 2005
24 x 24 x 1 inches (61 x 61 x 2.5 cm)
Earthenware tiles; electric fired, cone 03; laser
toner (glue transfer), glaze, cone 04;
scratch technique
Photo © artist

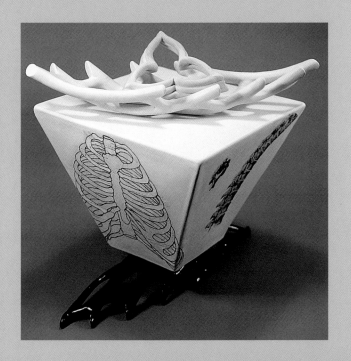

Top left: **Diane Eisenbach**
Untitled, 2004
11 x 13 x 9 inches (27.9 x 33 x 22.9 cm)
Hand-built stoneware and porcelain; electric fired,
cone 5; photocopy transfers, inlaid colored clay
Photo © artist

Top right: **Rebecca Robbins**
Arm-in-Arm, 2005
12 x 12 x 4½ inches (30.5 x 30.5 x 11.4 cm)
Slab-built clay; electric fired, cone 03; photocopy
transfer, silk screen, commercial decals, cone 017
Photo © Leonard Paul

Bottom: **Tracey Scheich**
Crouch, 2004
9 x 7½ x 3½ inches (22.9 x 19 x 8.9 cm)
Slab-built porcelain; electric fired, cone 6; digital
image on Lazertran
Photo © artist

Top left: **Erin B. Furimsky**
Nestle, 2004
6 x 9 x 6 inches (15.2 x 22.9 x 15.2 cm)
Hand-built stoneware, cone 6; electric fired in oxidation; commercial decal, cone 018; carved detail
Photo © Tyler Lotz

Bottom left: **Jennifer Mettlen Nolan**
Swing Shift: Heroine, Sweetheart, 2005
12 x 24 x 12 inches (30.5 x 61 x 30.5 cm)
Slab-built stoneware; electric fired, cone 05; photocopy decal
Photo © artist

Top: **Dan Anderson**
Pop-Truck Water-Tower Tea Set, 1997
9 x 12 x 8 inches (22.9 x 30.5 x 20.3 cm)
Hand-built stoneware with thrown parts; soda fired, cone 9; silk-screened decals, cone 017
Photo © Jeffrey Bruce

Right: **Elissa Armstrong**
Orange Lamb, 2004
8 x 5 x 2 inches (20.3 x 12.7 x 5.1 cm)
Hand-built earthenware; electric fired, cone 04, cone 018; decal
Photo © artist

Top left: **Paul Scott**
Scott's Cumbrian Blue(s), P/01/13/09/05, 2005
10 inches (25.4 cm) diameter
Earthenware plate; screen-printed, in-glaze decal collage (plate marked Keeling and Co, Warwick, 1790, Late Mayers)
Photo © artist

Top right: **Dalia Laučkaité-Jakimavičiené**
Animals Talking, 2003
13 inches (33 cm) diameter
Industrial porcelain plate; china paints, commercial decals, gold, platinum luster, cone 015
Photo © Vidmantas Ilčiukas

Left: **Juan Granados**
Workers: Carlos y Yo (Familia Series), 2005
17 x 17 x 2½ inches (43.2 x 43.2 x 6.4 cm)
Wheel-thrown stoneware; electric fired, cone 6; photo image transfer
Photo © Wesley Harvey

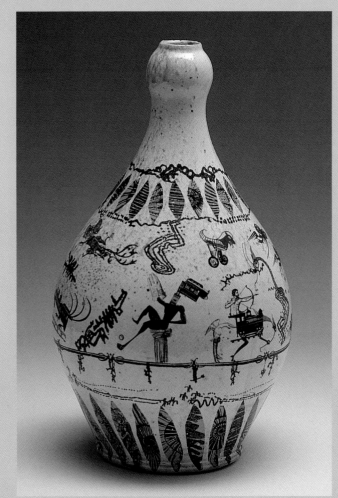

Top: **Barbara Hanselman**
What's in a Word, 2005
4½ x 12 x 12 inches (11.4 x 30.5 x 30.5 cm)
Pinch-strip-built stoneware clay; electric fired,
cone 5; photocopy transfer
Photo © artist

Right: **Carol Gentithes**
Post-Traumatic, Neo-Classic Vase, 2005
14 x 9 x 9 inches (35.6 x 22.9 x 22.9 cm)
Wheel-thrown stoneware; wood fired, salt
glazed, cone 10; silk-screened photo transfers, computerized transfers, cone 017
Photo © Tim Ayers

Springtime In Tim's Head

Top: **Christine Julin Shanks**
Extraction, 2004
15 x ⅛ x 11 inches (38.1 x 0.3 x 27.9 cm)
Slab-built porcelain; electric fired, cone 04; glazed, cone 06;
Lazertran, UV spray coat, semigloss protective coat
Photo © artist

Bottom left: **Lee Puffer**
Jack Pot, All You Wanted, 2005
22 x 10 x 10 inches (55.9 x 25.4 x 25.4 cm)
Hand-built earthenware; electric fired, cone 06; commercial decals,
cone 018; luster, cone 019; glass crystals
Photo © artist

Center right: **Brian Gillis**
On Gluttony: And the Reaffirmation of American Dreams, 2003
66 x 40 x 30 inches (167.6 x 101.6 x 76.2 cm)
Bathtub, found objects; unfired; stencil, cropped commercial
decals, unfired
Photo © artist

Right: **Eric Mirabito**
Angle Grinder, 2005
12 x 4 x 4 inches (30.5 x 10.2 x 10.2 cm)
Wheel-thrown and altered earthenware; majolica, electric fired; found object, silk-screened china paint decal, cone 017
Photo © Guy Nichols

Bottom left: **Rain Harris**
Bedazzle, 2004
10 x 7 x 7 inches (25.4 x 17.8 x 17.8 cm)
Slip-cast and altered porcelain, cone 06; decal, cone 012, multiple firings
Photo © John Carlano

Bottom right: **Lenny Dowhie**
The Elements of Tea, Historically Speaking, 1995
24 x 11 x 6 inches (61 x 27.9 x 15.2 cm)
Cast and constructed slipware; electric fired, cones 04, 06, 017; commercial slip, multifired; glazes, decal collage
Photo © artist
Courtesy of the Evansville Museum of Art

monoprints

Since early in the history of the graphic arts, ceramists have sought to adapt printmaking techniques to their art. In traditional print-making, a monoprint is made from an original design that was first created on a block or surface of some sort, whether stone, metal, or wood. (For convenience's sake, we call it a block even if it's a slab.) The block is inked, and a print is made by pressing a sheet of paper onto it. A set of two or more identical prints is called an *edition*.

Paul Andrew Wandless
Keeping the Star at Bay, 2005
14 x 14 x 2 inches (35.6 x 35.6 x 5.1 cm)
Low-fire clay, underglaze, underglaze watercolor; clay monotype from plaster slab

Similarly, a clay monoprint or monotype is a drawing or painting created with engobes, under-glazes, or slips—the typical "inks" of the ceramic world—on a flat surface such as plaster or paper, then transferred to a clay slab. To make an edition in clay, however, the image must be recreated before it can be transferred again to a new clay surface, so some variations between individual "prints" on clay are inevitable.

Monoprints and monotypes are printmaking terms that are often used interchangeably, but they have slightly different definitions. A *monoprint* utilizes printmaking processes such as screening and stenciling to create the image, though it can also incorporate hand-drawn elements. Variations from one print to the next sometimes create interesting and subtle changes. By contrast, a *monotype* is a ceramic drawing or painting created entirely by hand, without using any sort of printmaking tool at all. Such images are usually one-of-a-kind prints, since freehand work is difficult to reproduce exactly, though with a simpler design, edition-like multiples, sometimes called a *series*, are possible. My two pieces on the opposite page comprise a monoprint series.

To transfer an image onto a clay slab, you first mix and cast a plaster slab, then draw or screen print an image onto it with ceramic inks. The image is finally transferred from the plaster by pouring a slab made of casting slip onto it. The resulting monoprinted clay slab—a monotype or a monoprint, depending on how you created the image—can be a finished work, or it can be used with hump or slump molds (to make platters and plates), folded into vessel forms, or even used to hand-build sculpture.

Paul Andrew Wandless
Mallet…Bird…Struck #1, 2004
13½ x 13½ x 1 inches (34.3 x 34.3 x 2.5 cm)
Low-fire clay; underglaze, majolica, underglaze chalk; clay monotype from plaster slab, stamp
Photo © artist

Paul Andrew Wandless
Mallet…Bird…Struck #2, 2004
13½ x 13½ x 1 inches (34.3 x 34.3 x 2.5 cm)
Low-fire clay; underglaze, majolica; clay monotype from plaster slab
Photo © artist

Slips, Engobes, and Underglazes

You'll find monoprinting to be a versatile technique. Create the design with any colored slip, engobe, or underglaze that is suitable for the temperature range of the clay body you'll be using to print it on. Certain techniques (for example, screening or stenciling) require that you first thicken the ceramic ink so it won't bleed at the image's edges when it's printed. For drawing directly on surfaces with brushes, syringes, or rubber bulbs, no adjustment in viscosity is needed.

Slips and *engobes* are essentially water-thinned clay bodies with oxides, stains, or carbonates added for color. A fired engobe has a smoother surface because it has a little more flux in its base recipe than a colored slip, which has a rough, matte surface. For a smooth matte finish, use an underglaze, which is a mix of finely milled clays and colorants.

To add more body to the ceramic color, you can use a thickening medium, clear acrylic printing medium, or gum arabic, or you can let evaporation do the job by leaving the ink uncovered for a day. Most commercial colored slips and underglazes are just about the right viscosity, though some may require a bit of adjustment. Experiment with thickeners to determine the best consistency.

When making a slip from scratch, I use just enough water so it has the consistency of a thick yogurt, then add a little printing medium so it will have the firmness and the stickiness that are needed for printing. See pages 104–105 for slip and engobe recipes that are good for screening and stenciling; just add stains for color.

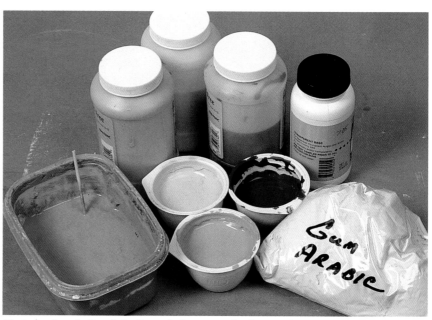

Clockwise from bottom left: slip made from scratch, pint containers of commercial underglaze, clear acrylic print medium, gum arabic, and small uncovered containers of underglaze

Bojidar Bonchev
Face 6, 2003
22½ x 14 x 4 inches (57.2 x 35.6 x 10.2 cm)
Plaster monoprint, cast porcelain; electric fired, cone 3; commercial decals, cone 016
Photo © artist

Plaster Slabs

Plaster is a versatile material for ceramists. Usually it's used to make molds, but a flat slab of plaster works great for image transfer processes. The slab must be smooth on one side and absorbent and porous enough so that a slip or underglaze will adhere well, yet remain dense enough to retain a crisp, clean image and stand up to repeated use.

A single company manufacturers all the plaster in the U.S. (over 30 types!), but the best results are obtained if you use pottery plaster #1 or #2. Each has a good balance of density and porosity, providing the requisite strength, durability, and surface needed for various transfer processes. See the pottery plaster guide on page 107 for more information.

Making a plaster slab in a mold box is easy and quick (it takes less than an hour), and you'll be able to use it for many years. Some of my slabs are up to five years old and still have excellent printing surfaces. Ultimately, the life of a slab depends on the quality of the plaster used and how efficiently the water and plaster were mixed together. After making a few slabs, you'll soon feel confident with this simple process. Work in a clean, ventilated workspace and plan ahead when possible: I recommend you let the plaster cure for 10 to 14 days before using it.

Constructing a Mold Box

L-shape mold boards are easily clamped together to make a square or rectangular box. I used pieces of pine that measured 1 x 4 x 16 inches (2.5 x 10.2 x 40.6 cm). Place a 1 x 1 x 4-inch-square (2.5 x 2.5 x 10.2 cm)

Bojidar Bonchev
Face 3, 2003
22½ x 14 x 4 inches (57.2 x 35.6 x 10.2 cm)
Cast porcelain; electric fired, cone 3; commercial decals, cone 016
Photo © artist

To make a mold box you'll need wood boards, square dowel rods, a power drill with a drill bit and a screwdriver bit (or a hammer), wood glue, wood screws or finishing nails, sandpaper, and polyurethane or shellac.

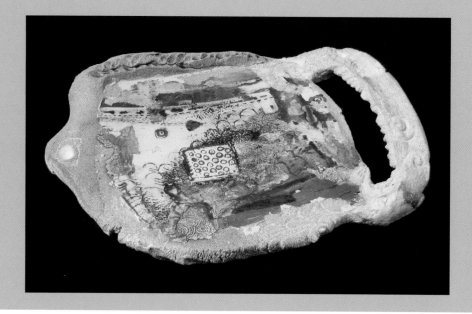

Sarah Rice
Weight of Giving **(series)**, 2005
5 x 15 x 15 inches (12.7 x 38.1 x 38.1 cm)
Hump-molded and hand-built raku; gas fired in
reduction, cone 6; earthenware slip monoprint
from plaster
Photo © artist

dowel rod across the end of a board, edges flush, and predrill a hole at each

end. Hold the wood firmly in place or use wood clamps to secure it to the tabletop. Slowly drill through the dowel rod but only slightly into the board itself (see photo 1). Predrilling keeps the wood from being split by the screws you'll insert later.

Apply a bead of wood glue across the end of the board and put the dowel rod in place (see photo 2). Use a power drill to insert the wood screws into the predrilled holes. (Alternatively, you could use a hammer and finishing nails; just be sure to make starter holes in the boards.) Use sandpaper to smooth all the surfaces. Waterproof the wood with a coat of polyurethane or shellac; the boards

will last longer and will be less likely to warp.

Now you're ready to make the mold box. Assemble the tools. Putting together a mold box only takes about 15 minutes. Start by positioning the four mold boards on a sheet of glass or clear acrylic (so the plaster will have a very flat, smooth surface). Butt the plain end of a board against the dowel end of another one and secure the joint of the L-shape with a C-clamp (see photo 3). Join the clamped L-shapes to each other (see photo 4). To ensure that the plaster slab will have an even thickness, check the level of the glass. Even it up with thin clay coils if necessary. Check the C-clamps to be sure there are no gaps between the boards. One clamp per corner should be fine, but use additional clamps for longer boards or if gaps still appear between the boards.

To prevent plaster from leaking at the seams, firmly press thin clay coils along the inside perimeters, then smooth them out (see photos 5 and 6). The finished plaster slab will have smooth, slightly rounded corners. If you prefer angled corners, put the clay coils outside the boards' seams instead.

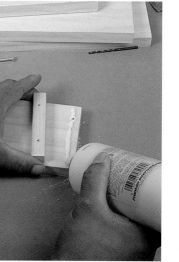

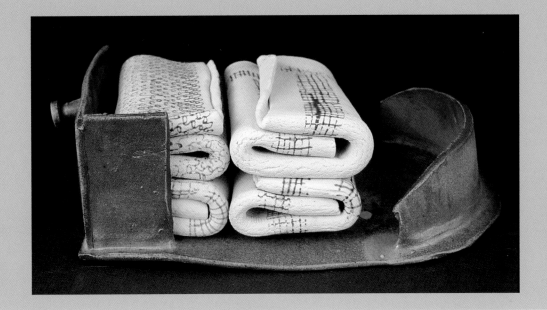

Kicki Masthem
Linen Drawers, 2002
5 x 14 x 8½ inches
(12.7 x 35.6 x 21.6 cm)
Linen made of paper clay,
stoneware drawer; linen: electric
fired, cone 04; drawer: soda fired
cone 11; paper transfer
Photo © artist

Use a brush to apply a thin, even coat of mold soap (see photo 7). The consistency of the soap should be very thin so as to prevent bubbles and brush marks. Check the directions on your brand for thinning instructions. Wait about 15 minutes so the soap can completely dry before pouring the plaster.

Calculating Volume

Figuring out how much plaster you'll need isn't an exact science, but a .7:1 ratio of water to plaster gives good results. Master mold makers simply estimate the volume and mix the plaster and water by sight and feel. Until you gain that kind of confidence, use this simple method to calculate how much plaster and water you'll need to make a plaster slab. Once the plaster has been mixed it sets into a solid mass pretty quickly, so you must make the right amount all at once.

Calculate the volume for the plaster slab by multiplying the interior dimensions of the box by the planned thickness (height) of the slab (see figure 1).

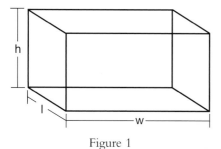

Figure 1

w x l x h = volume (in cubic inches)
14 x 14 x 1½ = 294

Divide the volume by 81 to find out how many quarts of water will be needed (use 77 to determine the number of liters). Round the answer up to the nearest whole number. Use the chart on page 107 to find the amount of plaster needed for that quantity of water. It's always better to have a little more plaster than not enough.

$$\frac{294}{81} = 3.6 \text{ (round the result to 4 quarts)}$$

From the chart, this amount of water, plus the plaster, will make 320 cubic inches (5,244 cc³). If you pour the slab to a thickness of 1½ inches (3.8 cm), you'll have 26 cubic inches (426 cc³) of plaster left in the bucket. Pour the slab a little thicker if you don't want any waste. Have another bucket of water nearby for cleaning your hands. There'll be more said about cleanup after you've cast the plaster. Just remember: Less water makes a harder but less absorbent slab, and too much water results in a softer, less durable slab.

Mixing Plaster

Weigh the cool, clean water and plaster into separate 5-gallon (18.9 L) plastic buckets. Wear a dust mask whenever you're handling dry plaster powder. The water's temperature affects how fast or slow the plaster thickens; warmer water speeds the process, so use cool water if you want to have a comfortable amount of time to work. Check the plaster for any lumps and break them up so the plaster can be mixed to a smooth consistency.

For making plaster you'll need two plastic buckets, a large-capacity scale, water, pottery plaster, and a timer.

Do your mixing only a short distance away from the mold box. Add the plaster to the water, making sure you distribute it evenly over the entire surface and taking no more than 30 seconds to do it (see photo 8). You'll see some islands of plaster bubbling as they settle below the surface (see photo 9). All the plaster must eventually be at the water's level or else sink below it.

Allow the plaster to *slake* for two minutes; start your timer as soon as you add the last bit of plaster. Slaking allows it to become fully saturated with water. For best results don't disturb the bucket during this time—any agitation will start the hardening process.

When the time is up, put your hand into the bucket and take a few seconds to do a quick check for clumps in the mixture. Position your open hand, palm down, about 1 inch (2.5 cm) from the bottom of the bucket as you rapidly rotate your hand at the wrist from side to side, as if waving. Use this motion while moving your arm from side to side and up and down in the plaster, but avoid stirring with your arm or moving your hand in circular patterns. Take extra care to avoid creating any air bubbles, which would leave pock marks on the slab's surface.

Mixing starts the hardening process. After just a few minutes, you'll notice a change in the plaster's thickness and temperature. Eventually the plaster attains a smooth, yogurt-like consistency; then it's ready to pour. The stages of thickening occur rapidly so start pouring the moment the plaster seems to be right. A few moments too late and it will pour like lumpy oatmeal, with a lot of air pockets to boot. Try the method I use for the thickness check: pull your hand out of the plaster with your fingers pointing down. If the plaster is thin enough to see your skin, it's not ready (see photo 10). When the plaster can cling and cover your hand like a thin glove, with little or no dripping, then it's time to pour (see photo 11). You'll also know it's ready when you can pull a subtle groove through the surface. Of course it takes practice to recognize the perfect moment. Just be ready to move fast.

9

10

11

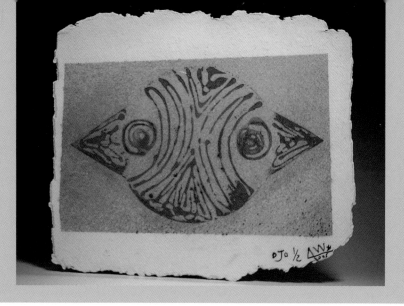

Casting the Slab

Starting in one corner, pour the plaster from the bucket, in an uninterrupted and steady flow, into the mold box. Let it spread across the bottom of the mold on its own, until it reaches the desired thickness (see photo 12). Pouring in one spot only helps the plaster attain an even consistency, and it will have fewer air pockets. Moving the bucket around too much while pouring creates those dreaded air bubbles. To dislodge any bubbles, "burp" the mold with the side of your fist (see

12

13

14

15

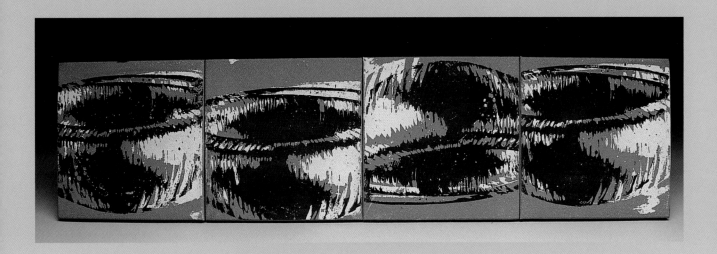

photo 13). You can use a long metal or plastic spreader to smooth the surface as it starts to set up.

Within 10 minutes the shiny surface will become dull and firm, and will feel warm to the touch. During the next 10 minutes, the plaster's temperature will increase enough to make the plaster sweat inside the mold. Now you can remove the mold boards and clay coils (see photo 14). Once the sweating starts, gently push on one side of mold to see if it will break free and slide over the glass's surface (see photo 15); if not, check again in a few minutes. You must remove the slab from the glass before this layer of moisture dissipates and creates a strong suction

between the two surfaces, making it very difficult to separate the two.

Sharp corners on the slab's edges tend to chip, so use a rasp (such as a Surform tool) or a putty knife to bevel all the edges slightly (see photo 16). Rounded edges also help the slab to dry evenly and make it easier to handle. Flip the slab so you can see how smooth your printing surface turned out. It should be completely free of irregularities. The plaster is still curing and hasn't reached its final hardness, so be careful not to scratch the surface. Let it dry for a minimum of 10 days (14 are better), raised off the table by two dowel rods. The longer you let the slab cure, the more durable it will be.

Cleanup & Tips

Plaster in a drain will harden and clog the pipes. Rinse your hands in a separate container of water first before washing them in a sink. Wait a couple of hours for any extra plaster in the bucket to set completely so you can easily clean it out in one piece. Just squeeze the sides of the bucket and the plaster should pop away from it. Always scrape and clean the mold boards with a putty knife so the next slab you pour will have nice smooth sides, too. If properly maintained the mold boards will last for years.

Making a Monotype

When I want to make a freehand painting or sketch, or create some spontaneous marks for a piece, I use a plaster slab as the canvas for my ceramic drawing. Later, I pour casting slip over the final image. The drawing is transferred when the slip absorbs the colored slip, engobe, or underglaze from the plaster. When the clay slab starts to stiffen, I pull it off the plaster; the (reversed) image comes away cleanly.

The first layers of color applied to the plaster will appear to be in the foreground and "behind" the ones applied later. This "drawing in reverse" takes a little practice—it's a different way of thinking. Of course if you apply colors next to each other rather than overlapping them, they'll appear in the final piece pretty much as you see them on the plaster.

You may already be familiar with drawing directly onto clay forms or slabs. When you work in that way, time is of the essence because the materials dry so quickly. Multiple sessions of drawing directly onto clay rehydrate it and can even weaken your clay form. On the other hand, plaster absorbs and holds the image until you're ready to pull it off—even if it's weeks or months later.

16

Ian Thomas
I've Lost My Sheep, 2005
12 x 12 x ½ inches (30.5 x 30.5 x 1.3 cm)
Slab-built terra cotta; electric fired, cone 04;
newsprint transfer, underglaze stick, cone 04
Photo © artist

Drawing on Plaster with Ceramic Ink

Drawing images with ceramic inks is simple and immediate. Gather together your prepared slips or underglazes and other tools. I used commercial colors for this demonstration, but colored slips made from recipes can be used as well. The greater the variety of ceramic inks, the more interesting the image can be. Before you begin, make sure the plaster slab's surface is clean so the slips and underglazes can adhere uniformly. Clean the drawing area by gently wiping the surface with a dampened soft cloth or sponge.

Underglazes and slips are fluid enough to be used as is. A wide variety of lines and marks can be achieved when you apply them with one of the tools from the assortment of rubber bulbs, brushes, and slip-trailing bottles that are available. Have several of these drawing instruments for each color so you won't have to stop too often to clean them before using a different color.

Assemble these articles for drawing on plaster (from left to right): commercial underglaze (pint size shown), slip trailers filled with water-thinned underglaze, rubber bulbs, trays of watercolor underglaze, and assorted brushes.

Paul Andrew Wandless
Good Boy, 2005
18 x 14 x 3 inches (45.7 x 35.6 x 7.6 cm)
Low-fire clay; underglaze, textural color,
metal; clay monotype from plaster slab, stamp
Photo © artist

Casting Slip

The simple outline drawing shown below was made with a slip-trailing syringe filled with water-thinned black underglaze (see photo 17). Unless you want the colors to bleed together, let the plaster absorb each color for a few minutes before applying another one.

To achieve greater complexity in the image, vary your application methods. I applied the next underglaze color by brush to give the lines a painterly feel (see photo 18). A transparent, washy look was achieved with a commercial watercolor underglaze (see photo 19). Put a soft cloth under your hand so you don't smear your work. When you've finished, let the colors dry for 15 to 20 minutes.

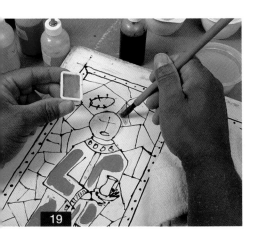

19

A ceramic-ink design drawn on a plaster surface is transferred by pouring *casting slip* over it to form a clay slab. The casting slip absorbs the image from the plaster as it dries. When the soft leather-hard cast slab is removed, the image comes off cleanly and is retained in the slab's top layer.

You have three options when choosing a casting slip. The first and easiest is to use a prepared commercial casting slip, already deflocculated (more about that in a minute). It's ready to mix and pour right away. Another choice is to thin a clay body with water to the consistency of cream, find the specific gravity to determine if the water-to-clay ratio is correct, and add a deflocculant for fluidity. The third way is to mix a casting slip using your own recipe (or choose one from the recipes on page 104) from dry materials, adding deflocculant to the water before mixing the slip to the proper specific gravity.

Measuring Specific Gravity

Specific gravity is the expression of the density of a *clay slurry* (finely divided ceramic materials mixed with water) as compared to an equal volume of plain water. For monoprint-

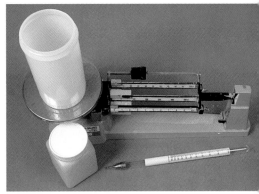

Items used to determine specific gravity are (from left to right): a plastic container to weigh water and slip, a triple beam scale, a hydrometer, and some deflocculant.

ing, a slip with a specific gravity between 1.80 and 1.84 provides good fluidity and a strong cast-clay slab. To measure specific gravity, place an empty graduated cylinder on a scale and adjust the reading to zero. Fill the cylinder with 100 milliliters of water and reweigh it; if the scale is accurate the water will weigh 100 grams. Empty and dry the cylinder. Refill it with 100 milliliters of the slip and note its weight. The weight of the slip divided by the weight of the water yields the slip's specific gravity.

$$\frac{182 \text{ mL}}{100 \text{ mL}} = 1.82 \text{ specific gravity}$$

A slip's fluidity is determined by its clay-to-water ratio. A specific gravity of more than 1.84 is too high: you can reduce the ratio by adding water to thin the slip. If the number is less than

Paul Andrew Wandless
St. Pablo, the Blacksmith, and the Potter, 2005
10½ x 14 x 2 inches (26.7 x 35.6 x 5.1 cm)
Low-fire clay, underglaze, underglaze watercolor;
clay monotype from plaster slab

1.8, let the slip sit overnight so it can separate a little, then skim some water from the top.

A quicker (though less accurate) way to measure specific gravity is with a hydrometer. This measuring instrument is a lead-weighted, vacuum-sealed tube with a specific-gravity scale inside it. To use it, gently release the hydrometer into a bucket of thoroughly mixed slip. The hydrometer must float freely (it will stand upright on its own) and won't touch the bottom if it's in a deep-enough container. When you first mix the slip, measure the specific gravity with the scale method just described; use the hydrometer to re-check it before subsequent uses.

Deflocculation

A *deflocculant* is an additive that evenly disperses clay particles in solution for a consistent, pourable liquid. Typically, low-iron porcelain and white stoneware have been the most popular clay bodies for slip casting because as liquids they're very fluid. The particles in iron-rich clay bodies, on the other hand, have a strong attraction to each other, which makes it difficult for an ordinary deflocculant to do its job. Fortunately, new polymer-type deflocculants now make it possible to slip-cast even with ferrous clays such as red earthenware.

Once you've mixed the slip to the consistency of cream, add a drop of liquid deflocculant; thicker slip (more like yogurt, say) may need more—up to three drops—added one at a time. Careful: too much deflocculant can cause problems, such as pockets of uneven drying and absorption in the clay slab. Over-deflocculation can be reversed by mixing single drops of a mild acid, such as vinegar, into the slip.

A deflocculated slip is also *thixotropic*, but its lovely fluid state will soon return to a yogurt consistency, so the slip must be stirred every day to keep it from getting too thick, or *viscous*. In a short time (a week or more), evaporation will likely affect the slip's specific gravity; check it with the hydrometer and adjust—most likely by adding water—if necessary.

Casting a Clay Slab

Now it's a simple matter of pouring casting slip over the image on the plaster. First, level the plaster slab with a bubble level and some shims. Use clay coils to build a dam around the perimeter of the slab's surface, brushing on some casting slip to secure the coils in place (see photo 20). If you want the clay slab to be a precise thickness, measure and mark a fill line on the clay wall. As you determine the slab's thickness, remember to allow for the shrinkage of your clay body.

Use a ladle or a pitcher to pour casting slip evenly over the image. (For a thin slab you can paint layers of casting slip over the image with a wide brush [see photo 21]). As you pour a deflocculated slip, you'll see how easily it flows and flattens out; a non-deflocculated

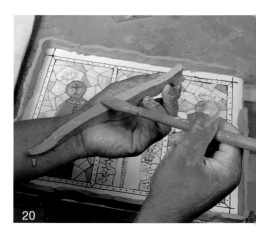

20

Kelly McKibben
Pitcher Set, 2004
11 x 20 x 9 inches (27.9 x 50.8 x 22.9 cm)
Wheel-thrown white earthenware; silk-screen
slip appliqué transfer, electric fired, cone 04;
clear glaze
Photo © artist

slip is less plastic and tends to hold its shape. Fortunately, both types have their creative uses. While the slip is drying, maintain its thickness by topping it off a little. If the slab is very large or thick, use a spreader to smooth the surface and coax out air bubbles (see photo 22).

After 15 to 20 minutes touch the surface to see if the slip is starting to stiffen. When your finger can't make a dent in the surface anymore, trim away the clay walls with a fettling knife so the clay slab can shrink easier, with less danger of cracking (see photo 23). After 30 minutes lift one

corner with a rubber rib to see if it will release from the plaster (see photo 24); if not, continue to check it every 15 minutes. When the clay has stiffened to the soft leather-hard stage, pull it up by the corners. For a larger slab, cover it with newspaper and a piece of cardboard or wood

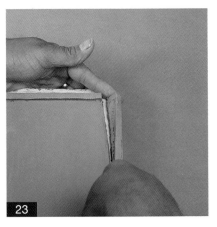

Screen Prints on Plaster

Another way to put an image onto a plaster slab is to screen print it there. Keep in mind the image will be reversed when it's pulled from the plaster surface. For some images this may not be a big deal, but if you're using text or numbers, you'll want to put a *reversed* image into the screen so that there aren't any surprises when you pull that first slab. Also, images that incorporate various areas of overlapping color must be screened onto the plaster with the same foreground-to-background consideration as mentioned earlier.

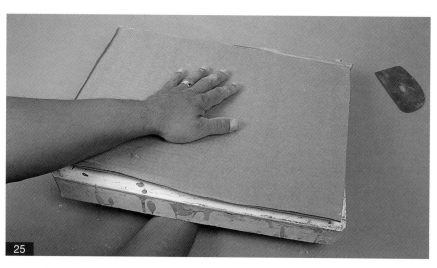

25

With minor adjustments, commercial colored underglazes are suitable for screen printing. You can thicken most of them by leaving them uncovered overnight or by adding a clear acrylic printing medium. Colored slips made from dry ingredients can also be used to further broaden your palette; see page 104 for recipes.

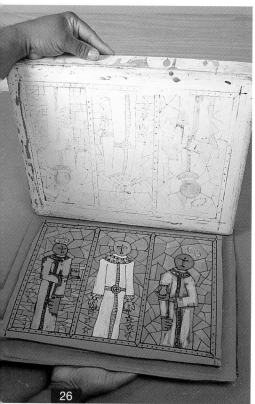

26

(see photo 25). Place one hand one top of the cardboard and one hand under the plaster slab, then flip them; lift off the plaster slab (see photo 26).

The monotype is done. You can build with it now or consider it a finished work (after firing, of course). If you plan on bending the slab into a different shape, do it as soon as it's strong enough to be handled but while it's still in the most malleable stage. Cut and handle a stiffer slab just like any other clay slab—this one just happens to have a printed image on one side.

Screening a Monoprint

You've already seen an image transferred with a light-sensitive emulsion screen, so for this demonstration I'll use screens with images created using stencil film and drawing fluid instead.

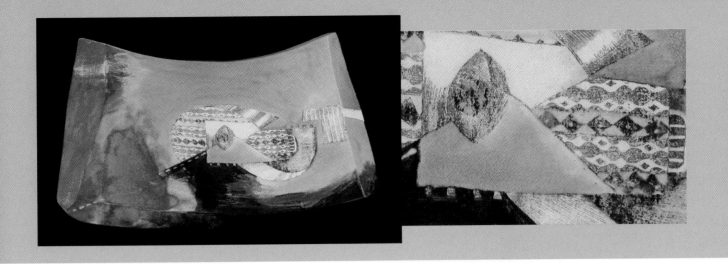

Elevate your plaster slab with square dowel rods so that circulating air can dissipate the moisture the slab absorbs from the drawing media. Put a little more ceramic ink than you think you'll need in a small container. For the best results, thoroughly mix it to a smooth consistency before you start printing. If it's too thin, the applied color will bleed at its edges; add a little thickener if that happens. Make a test print on a separate plaster slab or even at the bottom of the one you're about to use.

Place the first screen facedown on the plaster slab (remember, subsequent colors will overlap this one). Apply a bead of color (see photo 27). Use the squeegee to pull the color evenly across the image. Remove the screen (see photo 28). I wanted the chair to appear as if it were sitting in the foreground plane, on a wooden floor, so I printed the chair first. The floor is screened next, overlapping the chair's legs a bit (see photo 29). If I had wanted to add more detail or fill in large areas of color, I could have added some hand-drawn color to these printed elements. Once the image was completely dry, I used the same process for casting a slab and pulling the image as demonstrated on pages 68–69.

Top: **Ian Thomas**
Dinner for One, 2005
12 x 12 x ½ inches (30.5 x 30.5 x 1.3 cm)
Slab-built terra cotta; electric fired, cone 04;
newsprint slab transfer, underglaze paints and
sticks, glaze, cone 04
Photo © artist

Bottom: **Shalene Valenzuela**
Madeline, **from Watershed Picture Frame
series**, 2004
12 x 8 x 1 inches (30.5 x 20.3 x 2.5 cm)
Frame-molded ceramic; underglazes, underglaze
pencil, Watershed clay, electric fired, cone 04
Photo © artist

Drawings or Screen Prints on Paper

Imagery created on newspaper and transferred onto clay can be used for appliqué on sculpture or vessels. The pictures are drawn or screened (as demonstrated on plaster slabs, pages 66 and 71), then transferred onto clay using brushed, not poured, slip. This newspaper method is a great way to create small images or designs on paper-thin clay slabs that can later be collaged onto a leather-hard clay surface. Any clay body can be used for this process, but if you intend to appliqué or collage it onto other clay, use a slip made from the same clay to which it will be applied. Different clay bodies may have different shrinkage rates that would cause the collaged clay parts to pop off later as the piece is drying.

Making the Print

Use masking tape to secure a sheet of newspaper onto a wooden board or other flat surface. Lightly mist the newspaper until it darkens slightly from the moisture. Don't overdo it—spray just enough to dampen the newspaper slightly. Having a little moisture in the paper helps reduce the wrinkling that would otherwise happen when a slip or underglaze is applied. Some wrinkling is unavoidable with such thin clay slabs, but they smooth out when applied to a leather-hard clay surface.

As with other monoprinting processes, you'll need to create the image from foreground to background. The loose design shown in this demonstration was drawn with commercial under-

glazes in slip-trailing bottles (see photo 30). For appliqué or collage purposes, a simple design works well.

Let the slip image dry enough (the shiny surface will become dull) so

30

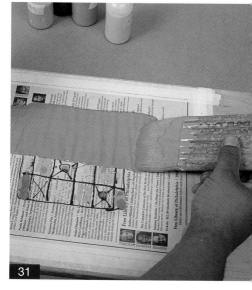

31

that the brushed-on slip won't smear it (see photo 31). Apply two to four coats (as thin as a dime or as thick as a quarter) with a wide brush; the backing slip only needs to be thick enough so it can be handled without tearing easily.

Smooth a sheet of dry newspaper over the brushed slip (see photo 32) and remove the masking tape. Flip the newspaper-sandwiched slab over so the image side faces up. To help the newspaper release cleanly from the image, lightly mist the paper first. Slowly peel off the damp newspaper (see photo 33). If the slab is thin, it will have wrinkles, so smooth these when you apply it to a leather-hard clay surface.

32

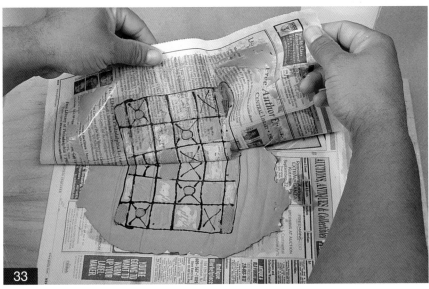

33

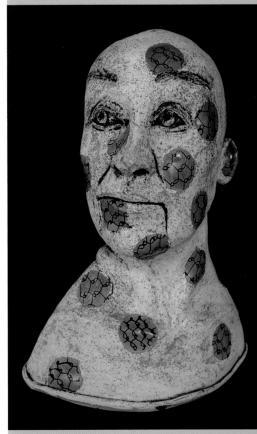

Kicki Masthem
Pensive, 2005
19 x 12 x 8 inches (48.3 x 30.5 x 20.3 cm)
Hand-built; electric fired, cone 04; paper transfer
Photo © artist

direct prints

Any time a picture is created right on the surface of the clay, it's called a *direct print*. The transfer of the image is immediate. The middleman is gone: there's no need for intermediary vehicles such as decal paper, newspaper, or plaster slabs. Direct printing uses portable and reusable image-making tools like stencils, screens, relief blocks, and stamps to make multiple images or designs, and they work with slips, underglazes, engobes, and glazes. Direct printing methods work great alone and of course can also be used in conjunction with any other process in this book.

Stephen Heywood
Double Lidded Serving Container, 2002
10 x 10 x 10 inches (25.4 x 25.4 x 25.4 cm)
Wheel-thrown stoneware; soda fired; stenciled, slip
Photo © artist

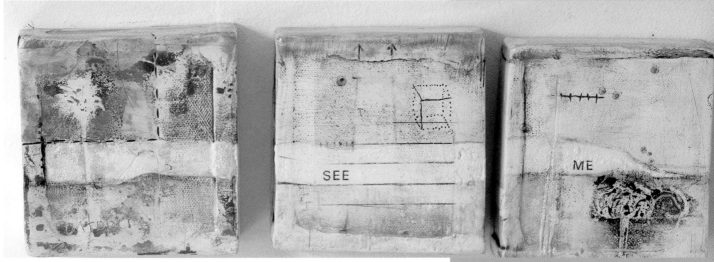

Jasna Sokolovic
See Me, Three Ceramic Blocks, 2005
4½ x 4½ x 1½ inches (11.4 x 11.4 x 3.8 cm) each
Slab-built white low-fire clay; electric fired, cone 06;
scratched, stamped, photo transfer
Photo © Nenad Stevanovic

Frank James Fisher
Real Teapot…Only .97¢!, 2005
6 x 6 x 2¼ inches (15.2 x 15.2 x 5.7 cm)
Slab-built porcelain; raku fired, cone 06; printing plate
impression
Photo © artist

Stencils

Stenciling is one of the oldest and easiest methods of transferring simple designs, text, or images to clay surfaces. We see stencils every day: they're used to spray-paint numbers, letters, and simple images onto cartons and crates, and in the work of graffiti artists.

The basic process is quick and direct: an image is cut into a thin sheet of stiff material such as plastic, cardboard, or metal. The stencil is affixed to a smooth clay surface, and color is then applied through the opening. You can purchase a wide variety of designs or hand-cut your own custom stencils to brush or spray slips, underglazes, and glazes of all temperature ranges onto any clay body.

Preparing the Stencil

Any stiff, flexible material that resists moisture is fine for a stencil. Paperboards can be used but they deteriorate quickly and are hard to clean. They'll work fine, though, for the short term. Transparency sheets, acetate, wax-treated paper, or commercial stencil sheets work best.

Most of the time you can draw the image onto the stencil material with a permanent marker or transfer it with tracing paper. For a digitally created design, use a computer to print it onto the appropriate type of transparency sheet, or photocopy (i.e. laser toner print) it onto acetate. Depending on the stencil material you've used, you'll cut out the image with scissors, a craft knife, or a commercial heat wand made especially for cutting stencils. Stencils aren't limited to simple pictures and silhouettes, but multipart images or those with continuous lines must be broken up somehow so the whole section doesn't fall away when you cut it. It's a little like creating a maze: the stencil material itself is the path that finds its way through the design, and the cutout areas are the maze's walls. You have to

Commercial stencils made of wax-covered paper (tan) and plastic (blue)

Far left: **Israel Davis**
Instructional/Play, 2003
17 x 18 x 1½ inches (43.2 x 45.7 x 3.8 cm)
Silk-screened wet slab-built terra cotta; slips, underglazes, glazes; electric fired, cone 04
Photo © Israel Davis

Left: **Meryl H. Ruth**
Paisley Pillow Purse, 2005
11½ x 10 x 6 inches (29.2 x 25.4 x 15.2 cm)
Slab-built with thrown spout; electric fired, cone 6; luster, cone 018; silk-screened
Photo © Robert Diamante

Right: **Joseph Pintz**
Half-Gallon Milk Bottles & Wire Crate, 2004
14 x 18½ x 10½ inches (35.6 x 47 x 26.7 cm)
Press-molded earthenware, found object; soda fired, cone 03; screen print, cone 015
Photo © artist

create a clear path around the walls for the maze to work. This is the approach I used to cut the suit stencil seen in the photo below.

Place a photo or drawing under a transparent sheet and use a marker to trace the contours of the forms. If the picture has a lot of detail, lay tracing paper over it (and under the transparent sheet) to help you focus only on the most prominent lines. Or create an outline image on a computer and print it directly onto a sheet of acetate

that's suitable for your printer. Wax-treated stencil paper can be drawn on directly, or you can trace the design onto it with carbon paper.

Cutting the Stencil

My favorite stencil-cutting tool—and the fastest—is the heat wand, which can be purchased at a hobby store (see photo 34). Its point is so small and the cutting action so smooth that you can obtain rounder lines and smaller details than with any other type of

cutting tool. Plug the wand in, set it on its special stand, and let the tip heat up for a minute. Be careful; the tip gets very hot. Practice cutting with the wand on a spare piece of stencil material. If you move too fast, it won't burn all the way through; too slow and it melts the edges. Cut on a piece of glass or wood or whatever surface is recommended by the manufacturer. You might use a craft knife instead of scissors, which are more likely to leave some jagged edges from the action of the blades. Just remember to always use a sharp instrument and make sure the surface under the stencil is smooth; a piece of wood or cardboard works well.

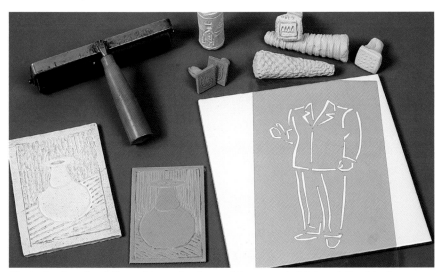

Clockwise from top left: a rubber roller for applying ceramic ink to printing blocks, a cylinder seal and assorted clay stamps to make impressions in soft clay, a custom stencil cut with a heat wand, and a linocut and claycut for relief printing and embossing

34

Charlie Cummings
Putting the Past in Its Place, 2005
6 x 6 x 1½ inches (15.2 x 15.2 x 3.8 cm)
Press-molded earthenware; electric fired, cone 1;
stenciled, dry-brushed slip
Photo © artist

Applying the Image

Stencils may be applied on fired and glazed surfaces as long the colorant is compatible with the glaze. New images can be layered over previously fired ones, as long as you fire the inks from the hottest to the coolest cone temperatures; for example, a cone 04 glaze may be laid over a cone 06 colored slip, but not vice versa.

The stencil must be secured to the clay surface if the edges of the image are to be clean. On soft leather-hard clay, just rub the stencil lightly to adhere it to the moist surface. For drier green ware, lightly mist the back of the stencil before laying it down. For bisque ware or a textured glazed surface, use a glue stick.

Choosing the right ceramic finish—whether slip or glaze—and application

method affects the look of the fired stencil's surface. A variety of visual effects are possible by brushing or spraying a slip, underglaze, or glaze through a stencil. The method you choose ought to be based on the qualities of the

35

36

Clockwise from top left: commercial underglazes, a container of slip, a handheld sprayer of water-thinned underglaze, two custom-made stencils, and a commercial stencil

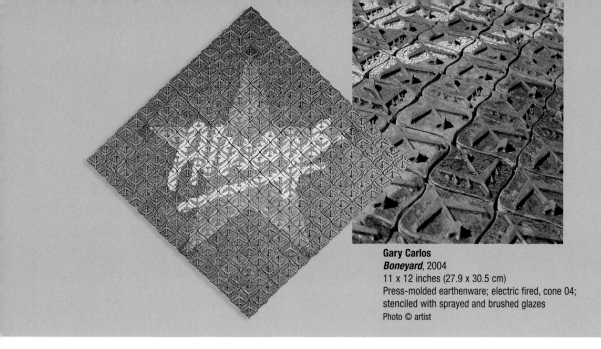

Gary Carlos
Boneyard, 2004
11 x 12 inches (27.9 x 30.5 cm)
Press-molded earthenware; electric fired, cone 04;
stenciled with sprayed and brushed glazes
Photo © artist

receiving surface and what sort of texture you'd like to create.

Stencil slips and underglazes onto green ware, or use glazes on bisque ware. Adjust the thickness of the ink to achieve various textures (testing always helps!). While cream-like finishes brush on smoothly, the yogurt-like consistency of a slip or a glaze will leave the brush's texture on it. A handmade slip, thickly brushed onto the blue stars in photo 35, has deep texture in it, while the "TV head" images are smoother from creamy commercial underglazes.

No matter what application method you use, as a rule of thumb, the finish is sufficient if you can no longer see the surface through the colorant. Gently and smoothly peel the stencil away, using both hands if necessary so it won't shift and smudge the edges (see photo 36).

A handheld sprayer or pneumatic spray gun creates the smoothest, most even, and crispest image. The handheld type (available in most hardware stores) has a removable jar you can fill with water-thinned slip or glaze, but it can

only be applied to a vertical surface. For flat surfaces, the spray gun does an even better job, though you'll also need some type of spray booth. Be aware that no matter which type of sprayer you use, if the colorant isn't thinned enough it will clog the nozzle.

Secure the stencil firmly to the clay as you spray the underglaze evenly in its open areas. A stenciled image stands out even more clearly on the clay if it's layered over a colored background. In photo 37 the "TV-head" image, sprayed on in black underglaze, looks even sharper when layered over brushed-on yellow and light blue underglazes than on terracotta clay alone.

Everything cleans up with water. Your plastic and wax-coated paper stencils are reusable, so have fun experimenting with them. In time you'll have a collection of interesting shapes that can be used in new ways and combinations on future projects.

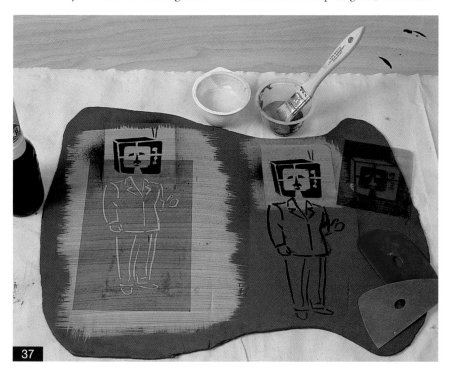

37

Scott Rench
China Stone, 2004
7 x 12 x 1¼ inches (17.8 x 30.5 x 3.2 cm)
Stoneware; airbrushed and printed underglaze,
electric fired, cone 03; hand silk-screened
Photo © Smitty Image

Screening Underglaze

Direct screening is done on a clay slab shortly after rolling it out. Applying the ink at a later drying stage risks poor fit and cracking. Be sure that the ink's firing temperature is compatible with the clay.

Roll out a clay slab large enough to accommodate the screened image. Trim the slab edges and smooth the surface with a rib. Place the slab on a worktable covered in canvas or newspaper, and position the screen on the clay. The first screen I used here is a digitally manipulated image that was burned into a light-sensitive emulsion.

Pour a bead of underglaze the consistency of honey across the screen. If the underglaze is too thin, it will bleed at the edges of the screened image. Pull the print with a squeegee (see photo 38), then remove the screen and check

38

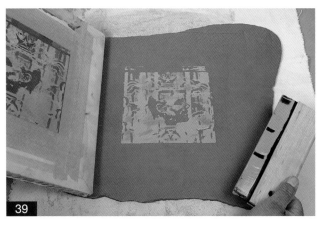

39

40

Anna Velkoff Freeman
Half Full or Half Empty?, 2005
Left, 3¼ x 3½ x 3½ inches (8.3 x 8.9 x 8.9 cm)
Right, 3½ x 3½ x 3½ inches (8.9 x 8.9 x 8.9 cm)
Thrown and altered porcelain; electric fired, cone 6;
linocuts, underglaze
Photo © Richard Nickel

Relief Block Prints

the image for color and quality (see photo 39). Some underglazes appear pale or somewhat transparent when printed, but they darken when dry. (The screen shown here has a lot of open areas and halftones that capture the different levels of opacity in the original, and you can see that the screened ink is also of varying thickness.) If you don't like the result just use a rib to scrape off the unwanted image and screen print a new one.

Layering various pictorial elements is simple to do. The face and text in photo 40 were created in the screen with drawing fluid and screen filler, and screened with a cone 06 commercial colored slip. After 10 minutes or so, the first underglaze had dulled and dried enough to print the suit image from a hand-cut stencil with another cone 06 slip. After the piece is fired, I can stencil or screen more images with different ceramic finishes and refire them at lower cone temperatures. I prefer to use subsequent finishes that must be fired two cones cooler than those previously applied so they won't burn out the already-fired designs.

Relief prints are made by pressing the raised areas of any found object or material, coated ("inked") with ceramic color, into soft clay. (The relief process, by the way, reverses your design's image or text.) Some typical relief-printing methods used in ceramics include block printing, stamping, and embossing. The design may be executed in a variety of materials, such as linoleum, clay, plaster, wood, rubber, fabric, or from found objects.

A design is drawn or traced onto block material, and all the negative areas (those you don't want to print) are carved away with gouges, knives, and chisels—or even with motorized engraving tools. The raised surface that remains is a *relief image* (whether a linocut, woodcut, or claycut) that will be inked with underglaze or slip and printed directly onto a clay surface. Naturally, the physical characteristics of the block material affect the look of the print. Each one yields a unique line quality, even though the same tools are used for all. An uninked block, pressed into soft clay, is an *embossing*. A stamped image is simply a smaller version of an embossing.

Woodcuts

Choose from many kinds of wood for a woodcut: softwoods like pine, balsa wood, basswood, poplar, and lauan are easy to cut. They can be found, as planks or blocks, in any lumber supply store. Woodcuts inked and printed (or just embossed) into clay retain the natural grain of the wood; each species has a distinctive pattern, so choose the one that best suits the image you have in mind. Unless you want to incorporate whatever marks are already there, sand the wood surface with fine-grit sandpaper first. As much as possible, carve along, not against, the wood grain. Cutting across the grain is okay, but the wood is likely to chip where you don't want it to. Motorized carving tools, with their wide variety of attachments, solve that problem, and they work nicely along with more traditional gouges.

Linocuts

Linoleum (or simply "lino") is a thin, flexible material with a smooth gray or brown surface and a backing made from hessian, a coarse fabric. Linoleum is great for carving any type of line or shape. It's completely responsive to sharp tools, and the backing prevents

Earline M. Green
Guardian of Comfort, 2000
20 x 10 x 9½ inches (50.8 x 25.4 x 24.1 cm)
Wheel-thrown and slab-built quilt stoneware;
electric fired, cone 5; bisque, cone 1; underglaze and glaze, cone 5; linoleum embossing
with stamped underglazed patterns
Photo © Harrison Evans

chipping. Lightly sand the surface with fine-grit sandpaper if it has any scratches or imperfections. Sanding also removes any oils in the surface, so it will be easier to ink with underglaze or slip. In fact, new linoleum some-times has a surface so smooth and tight that underglaze might not adhere well to it. To more easily carve an old, stiff sheet of lino, soften it up by holding a blow dryer or heat gun 6 to 12 inches (15.2 to 30.5 cm) from the back for a few minutes. Move the heat source from side to side rather than concen-trating it on one spot. An unmounted lino sheet can be too flexible, so you might want to glue it to a wooden board, or buy the mounted variety.

Claycuts

Clay's soft, responsive surface allows for an even wider variety of marks and impressions than are possible with a hard surface such as wood or linoleum. Slabs (approximately 1 inch [2.5 cm] thick) are easy to carve and emboss, so you'll have a lot of options with the types of lines and shapes that can be used for the image. Use the same gouges for lino and wood or try a ceramic loop or carving tool. The slab can be smoothed with a wet rib or lightly textured to complement the image in a manner similar to the way that wood grain does with woodcuts. But, a cut clay block must be dried slowly and carefully if you are to avoid warping it. Bisque fire the clay before you print with it.

Making a Relief Block

The cutting tools and carving process are basically the same for wood planks and blocks, linoleum sheets, and clay slabs. Buy them at a local hobby or art supply store. Choose a sturdy table or surface to work on. Take care not to carve with your hand in the path of the cutting tool—don't hold the block on your lap either. Gouges are very sharp!

Start off by putting a guide image onto the block. Keep in mind that the final printed image will be reversed. Draw directly on the surface or trace the image on with carbon or tracing paper. To easily reverse letters or num-bers, print them first on a sheet of acetate, then flip it over onto the receiving surface and trace them with carbon paper—you'll have perfectly reversed characters. Computer software can easily "flop" an image that you then print onto paper or acetate.

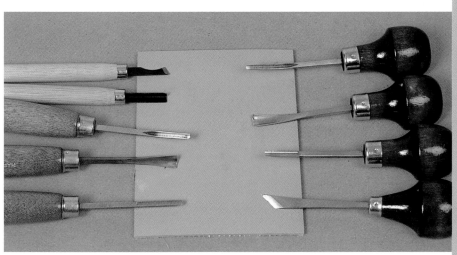

Gouges for carving linocuts

Prepare the wood and lino surfaces for transfer by lightly sanding them; prepare the clay by smoothing it with a rib. Secure the image and the carbon paper to the wood or lino with masking tape across the top, then use a pencil to follow the lines with enough pressure so a carbon image will be transferred to the surface beneath. An alternative to carbon paper is tracing paper, though it will reverse your image as you draw it. Draw on the tracing paper with a soft graphite pencil that makes a nice dark mark. Flip over the tracing paper and secure it with masking tape, then rub the paper or retrace the image. Moisture can hold paper to a clay block, so just use light pressure with a dull pencil to follow the lines of the original image (see photo 41).

Meryl H. Ruth
Turquoise Tea Tote, 2005
11 x 11 x 6 inches (27.9 x 27.9 x 15.2 cm)
Hand-built and slab-built stoneware and porcelain; electric fired, cone 6; lusters, cone 018; embossed
Photo © Robert Diamante

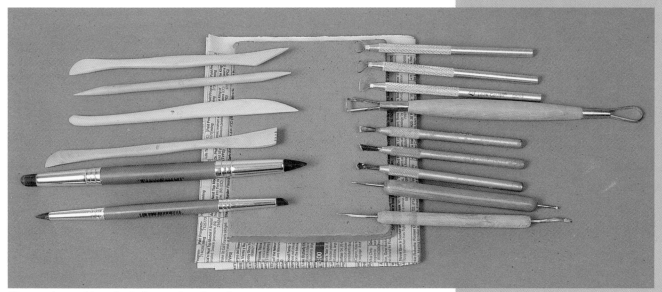

Clay carving and loop tools for carving claycuts

Left: **Douglas E. Gray**
Bibek's Gift, 2003
18 x 28 x 2 inches (45.7 x 71.1 x 5.1 cm)
Porcelain tiles; electric fired, cone 6; embossed digital image
Photo © artist

Right: **Linda Doherty**
Mug, 2005
4 x 2½ x 2¼ inches (10.2 x 6.4 x 5.7 cm)
Extruded and hand-built stoneware; gas fired in reduction, cone 10; texture transferred by rolling clay cylinder on textured paper
Photo © artist

Far right: **Paul Andrew Wandless**
V.O.A.V., #1 and #2, 2005
6 x 10 x 7 inches (15.2 x 25.4 x 17.8 cm)
Hand-built low-temperature clay; (left) linocut print, (right) woodblock print

Carving the Block

Practice with the gouges to get a feel for the cutting pressure needed to cut with them and the best angle to hold them. Experiment with all of them—they're really versatile. Remember to move the tool away from your body as you cut and keep your hand out of the cutting path.

Put pads of folded newspaper or cardboard between a wood or lino block (see photo 42). Clamps let you position the block at a variety of angles for greater control, or you

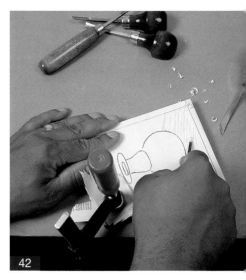

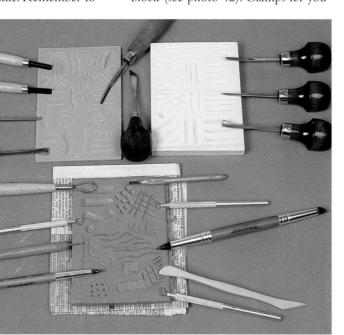

Each type of gouge can carve a different width and depth of line, and of course the block material itself contributes to the line's quality. Various cutting tools and some representative marks are shown here on (clockwise from bottom) clay, linoleum, and wood.

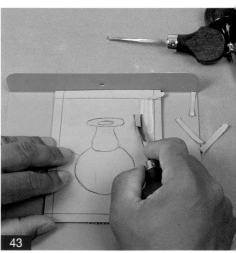

might use a bench hook instead, as shown in photo 43. (Though you can buy one, it's simple to make one by attaching a square dowel rod on opposite sides at each end of a piece of wood.) Push one edge of the block against the back wall of the bench hook. Carving a clay block, on the other hand, doesn't require much force. Put a piece of canvas or newspaper under the clay to keep it in place (see photo 44).

Using the drawing as a guide, start carving away all the areas that you don't want printed. The remaining raised areas of the block will be the printed image.

Preparing the Ink

Regardless of the type material you carve, you'll utilize the same basic process to print the images. For example, the linocut in photo 45 (at left) was used to create an ink-printed (and unfired) clay slab (center left), an embossed slab (center right), and a finished embossing on which glaze had been applied then wiped off the raised surface at bisque stage. Use any clay body that's appropriate for the firing temperatures of your slips and underglazes. A firm slab is best for receiving an ink printing and a soft one better for embossing; plan on making some test prints, too.

The key to creating a successful relief print is getting the proper consistency in the ceramic color. Of all the transfer techniques, it's relief printing that requires the thickest and stickiest consistency of slip or underglaze. When the color is too thin the block will be flooded with color and the resulting prints blurred. A too-thick consistency makes it hard to ink up the block, and the prints will be splotchy. You want it to be like hair gel or honey—sticky enough to adhere to the rubber roller but not thin enough that it can drip into the block's carved areas.

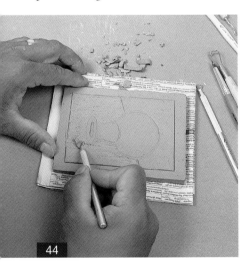

44

45

85

Paul Andrew Wandless
Linopots, 2005
Approximately 2 x 4 inches (5.1 x 10.2 cm) each
Low-temp clay; linocut embossed tiles, low-temp glaze and underglaze

Commercial underglazes and slips are too thin to use right out of the container. Either leave the containers open overnight to thicken, or add clear acrylic printing medium, which also helps increase stickiness. Colored slips can also be made from dry materials and adjusted to proper printing consistency; for some recipes, see the slips and clays recipes on page 104. The super-easy option is to purchase semimoist commercial underglazes; they're ready to use right out of the tube.

Block Printing

I'll demonstrate how to print a linocut with semimoist underglaze from a tube (see photo 46). Squeeze a bead of underglaze roughly the width of the block onto the acrylic sheet. Evenly spread the underglaze over the sheet by rolling it out, back and forth and side to side (see photo 47). Once the roller has an even coat of color, roll it lightly over the relief block; heavier pressure will put unwanted color into recessed areas (see photo 48). Once the

block is inked, put the roller back onto the acrylic.

Immediately turn the inked block facedown onto the clay slab. Don't wait too long or the underglaze will start to dry and it won't transfer as cleanly. Firmly but gently rub the back of the block with a mini-roller, a wooden spoon, or your hand in a circular motion (see photo 49). Hold two corners of the block and slowly pull it away from the slab (see photo 50). Examine

46

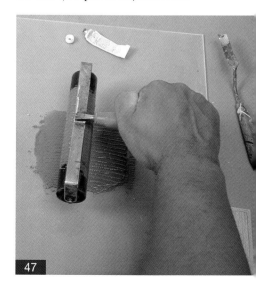

47

Peter Castle
Back to the Plate, 2004
9½ x 9½ x 3⅛ inches (24 x 24 x 8 cm)
Slip-cast earthenware; silk-screen decal resist; electric fired, cone 05;
sawdust fired
Photo © Malcom Bennett

Chris Stanley
Da Vinci Meets Disney, 1999
30 x 30 x 4 inches (76.2 x 76.2 x 10.2 cm)
Slip-cast porcelain; hand-drawn cut paper stencils, engobe, cones 10 and 11
Photo © Eric R. Johnson

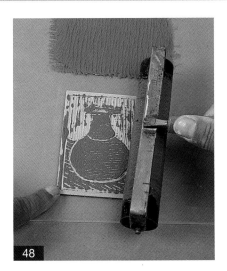

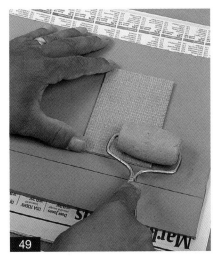

your first print to see if it has enough ink. Was the pressure right?

Use a palette knife to gather up any unused underglaze; store it for later use. Spray the block with water and wipe it clean with a rag (see photo 51). Don't soak the surface. Use just enough water to soften the underglaze so it'll wipe off easily. Clean the roller and acrylic, wiping first with newspaper, then dry it with a rag. Your printed slab can be used for any kind of hand-building project.

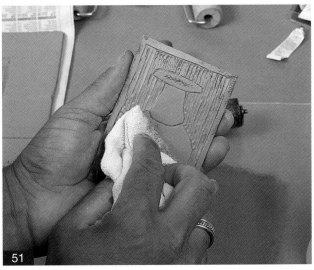

Block Embossing

The same relief blocks, uninked, can also emboss clay, like a large stamp. You can then bisque fire the impression and add some glaze color to the recessed areas. When I rolled the commercial product onto an acrylic sheet, I found it to be too thin and not sticky enough either, so I adjusted its consistency with clear acrylic printing medium, mixing well (see photo 52). The underglaze should be so thick that it can barely drip, and once it does, it should hold its shape well (see photo 53).

Load a foam roller with underglaze (see photo 54). Ink the bisque-fired claycut on a sheet of newspaper (see photo 55). Lay a soft clay slab on the clay cut and apply even pressure with

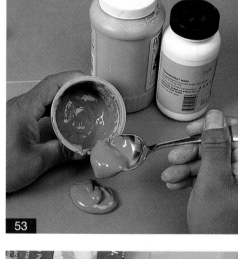

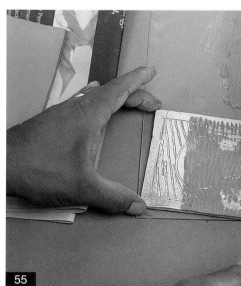

Vanessa Grubbs
American Beauty Rose Holder, 2004
12 x 5 x 5 inches (30.5 x 12.7 x 12.7 cm)
Thrown and altered white earthenware;
electric fired, cone 04; rubber stamped
Photo © Eddie Ing

a mini-roller to transfer the image. Pull the clay away from the block (see photo 56). Clean your tools with water when you're finished.

To add color to a bisque-fired embossing, apply glaze, underglaze, or stain in the recessed areas with a brush or other applicator and let it air dry. Use a damp cloth or sponge to wipe off any color from the high areas (see photo 57). Fire the piece to the appropriate cone temperature.

56

Amy Sanders
Cut Vase, 2004
8 x 6 x 3½ inches (20.3 x 15.2 x 8.9 cm)
Thrown, cut, and reassembled stoneware; electric fired, cone 6; impressed handmade clay stamps
Photo © Shane Baskin

57

Left: **Paul Andrew Wandless**
Red Chair, 2005
5 x 5 x 2 inches (12.7 x 12.7 x 5.1 cm)
Thrown low-temperature clay; glaze, cone 04;
hand-carved, stamped

Right: **Steve Irvine**
Double Wall Bowl, 2003
2⅞ x 11 x 11 inches (7.3 x 27.9 x 27.9 cm)
Wheel-thrown stoneware; reduction fired,
cone 10; stamped and carved texture; post-firing
23-karat gold leaf
Photo © artist

Far right: **Hodaka Hasebe**
Untitled, 2005
12 x 12 x 2 inches (30.5 x 30.5 x 5.1 cm)
Wheel-thrown white stoneware; gas fired in
reduction, cone 10; applied stamped clay
Photo © artist

Stamps and Cylinder Seals

The different types of stamps that can be used to make impressions in clay include set type, clay stamps, batik stamps, commercial rubber stamps, found objects, and cylinder seals.

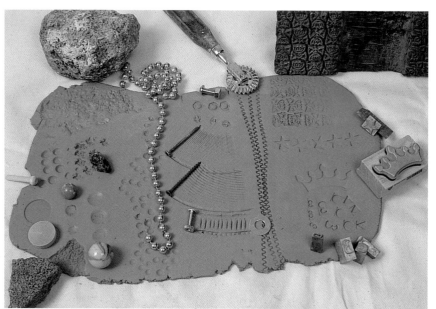

Found objects, commercial stamps, and the impressions they create

Stamping is probably the oldest of all relief techniques. It's been used in ceramics by sculptors and potters alike for thousands of years. Hand-carved and commercial stamps come in limitless varieties of shapes, sizes, and materials. Here you'll learn how to make several different types of stamps.

Most stamps are the size of your palm or smaller, so you can simply push them into the clay—a small version of embossing, really. Stamp designs are simple—usually just a single shape, design, texture, number, or letter. They're meant to be minimal—don't think of them as miniature printing blocks. Relief blocks are usually larger and more complex, with many visual elements. Another style of stamp is the rolling stamp, or *cylinder seal*. Such a tool creates a band, or *frieze*, in the clay. It's a great way to repeat a visual element or pattern over and over.

Found objects, rocks, hardware, coarse fabrics, and just about anything else with a raised surface can also be used to make impressions and create unique designs. Try carving plaster or wood stamps with the same tools used for making relief blocks.

Ball and Cone Stamps

The quickest and simplest stamp needs no tool at all. Roll a little soft clay into a ball or a coil about the thickness of your finger and flatten one end. Push the flattened end into an object or surface—concrete, maybe, or a shell—and

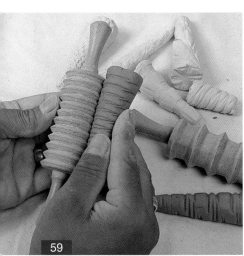

58

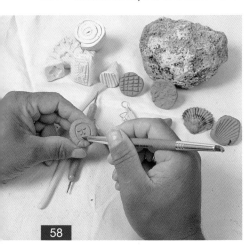

59

you're done. Carving is another option (see photo 58).

For a slightly larger stamp, roll a larger, fatter coil into a carrot shape and flatten the fat end. Carve the cone or roll it along a texture (see photo 59). Mark the flat end, too, for two stamps in one. If you use the narrow end as a pivot point, you can make circular patterns.

Bisque Stamp Templates

For a design with a particular orientation, such as a letter or number, make a bisque clay template. Roll out a clay slab and cut it into a square or rectangle (for 90° corners, use a carpenter's square or ruler). Draw a grid into the surface with a shaping tool. Create the designs in the open spaces with

60

shaping and carving tools (see photo 60). Let the template dry slowly, so it won't warp, and bisque fire it.

Use a bisque template to make stamps in a couple of different ways. You can push a small slab into the template (these will be reversed impressions; see photo 61), then trim around the design with a fettling knife and score the back. Attach the piece with joining slip to a scored and flattened clay ball or coil (see photo 62). An even simpler

61

62

91

Left: **Jacqueline Kierans**
In the Woods, 5 p.m., 2003
12 x 12 x 3 inches (30.5 x 30.5 x 7.6 cm)
Slab-built stoneware; electric fired, cone 6; silk-screened in-glaze decals, silk-screened over-glaze decals, cone 017
Photo © Gregory Staley

Right: **Earline M. Green**
Milestones of Rags, Polyester and Clay, 2000
15 x 18 x 15 inches (38.1 x 45.7 x 38.1 cm)
Wheel-thrown and slab-built quilt stoneware; electric fired, cone 5; bisque, cone 1; underglaze and glaze, cone 5; linoleum embossing with stamped underglaze
Photo © Harrison Evans

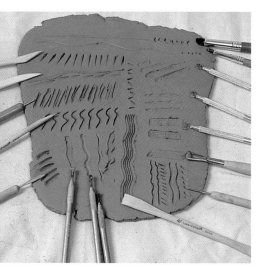

Clay carving and loop tools, and the lines they create

way is to push a ball of clay over the individual design you want. Bisque these stamps; the stamped design will face the right way when it's pushed into a soft clay surface.

Cylinder Seals

A cylinder seal is a rolling pattern-maker that can impress simple patterns or complex narratives into clay. Find a cylindrical form of the diameter you need. Cut a paper template of it and devise a picture to fit the space. Be sure to take advantage of the full range of your carving and shaping tools.

I rolled out a slab of clay and trimmed it to size with the paper template (see photo 63). A dull pencil or a clay shaping tool impressed the design into the clay as I traced the design's lines. Remove the paper back when finished and touch up any areas that need further attention (see photo 64).

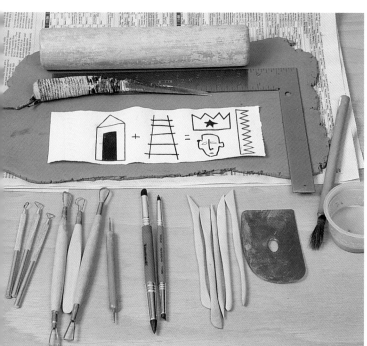

The materials for making a cylinder seal include (on the newspaper, top to bottom) a rolling pin, a fettling knife, a soft clay slab with the image to be transferred placed on it, and a carpenter's square. Below them are (left to right) loop and clay-carving tools, and joining slip and a brush.

63

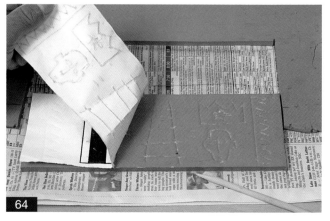

64

Wrap newspaper around the cylinder to prevent the clay from sticking to it. Score and slip the edges of the slab, then wrap it around the cylinder (see photo 65). Lightly smooth the seam with a soft rib or with your fingers (see photo 66). Now remove the clay tube from the cylinder and cap both ends with cut, scored, and slipped pieces of clay (see photo 67). Let the clay stiffen to the leather-hard stage, then carve the image; the lines will have nice clean edges when made with stiffer rather than softer clay. Follow the design impressed into the surface (see photo 68). Of course you can also draw a freehand design directly on the clay cynlinder.

After the image is carved, poke vent holes in the caps. Allow the clay to dry slowly. With larger vents a dowel rod can pass through them so the seal has handles, like a rolling pin. Once the cylinder is bisque fired, it's ready to roll (see photo 69). This design tool will last for many years.

Mishima

While stamped images alone are fun to work with, color can be added to the impressed areas, a technique known as *mishima*. When the stamped piece is leather hard, brush or trail underglaze or colored slip into the impressed areas. Let the color stiffen a little, then use a soft rib to wipe it off the high spots. To add color to bisque ware, use glaze instead.

65

66

67

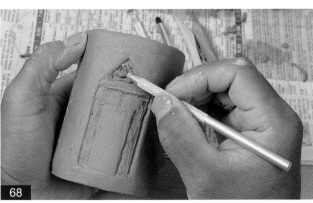

68

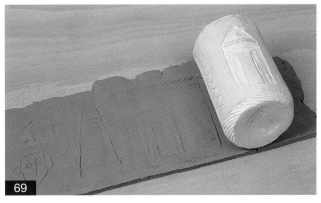

69

Jason Kiley
Pin-Up Cups, 2005
4 x 3 x 3 inches (10.2 x 7.6 x 7.6 cm) each
Thrown porcelain; gas fired, cone 10; spray stencil
Photo © artist

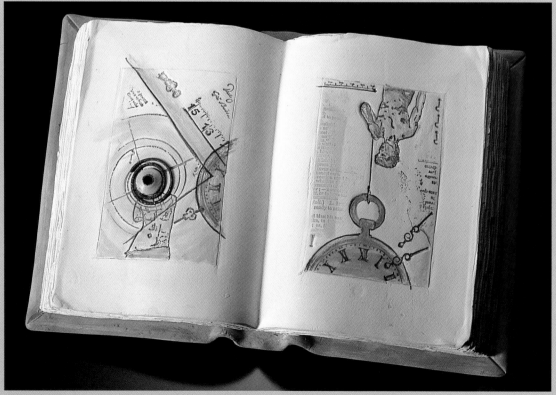

Abby Huntoon
Hang Time, 2005
5½ x 8 x 1 inches (14 x 20.3 x 2.5 cm)
Mold-pressed, slab back earthenware; electric fired, cone 05; commercial stamps, underglaze, acrylic paint, cone 4
Photo © Robert Diamante

Right: **Kirsten Jacobsen**
And Then, Double Jar, 2005
8 x 8 x 5 inches (20.3 x 20.3 x 12.7 cm)
Slab-built earthenware with terra sigillata; electric
fired, cone 04; stenciled, stamped
Photo © Tony Deck

Bottom left: **Scott Rench**
Place, 1996
27 x 24 x 2 inches (68.6 x 61 x 5.1 cm)
Brent slab roller stoneware; airbrushed, screen-
printed, underglaze, slumped, cone 04
Photo © Smitty Image

Bottom right: **Frank James Fisher**
Hi-Protein Drink, 2005
8½ x 5½ x 1¼ inches (21.6 x 14 x 3.2 cm)
Slab-built porcelain; raku fired, cone 06; printing
plate impression
Photo © artist

Left: Stephen Heywood
Teapot with Sugar Shaker Lid, 2002
6 x 9 x 4 inches (15.2 x 22.9 x 10.2 cm)
Wheel-thrown, soda-fired stoneware;
slip stencil number
Photo © artist

Bottom left: Frank James Fisher
Some Tea and the Latest News, 2004
7½ x 8 x 8 inches (19 x 20.3 x 20.3 cm)
Slab-built porcelain, wheel-thrown parts; raku fired,
cone 06; printing-plate impression
Photo © artist

Bottom right: Meryl H. Ruth
Woven Canteen Vessel, 2005
17 x 12 x 5 inches (43.2 x 30.5 x 12.7 cm)
Slab-built stoneware with thrown and altered parts;
electric fired, cone 6; lusters, cone 018; embossed
and airbrushed lace
Photo © Robert Diamante

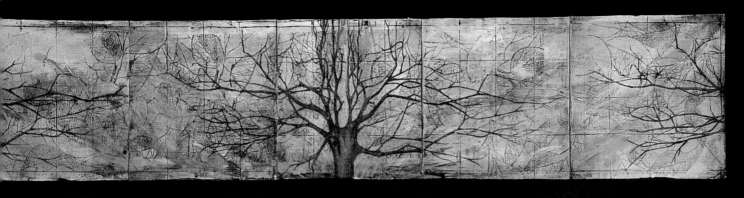

Jacqueline Kierans
Untitled #2, 2005
9 x 45 x 1 inches (22.9 x 114.3 x 2.5 cm)
Slab-built stoneware; electric fired, cone 6; screen-printed, cone 04, in-glaze silk-screened decal
Photo © Gregory R. Staley

Peter Castle
Back to the Plate, 2004
9½ x 9½ x 3¼ inches (24 x 24 x 8.3 cm)
Silk-screen decal resist; electric fired, cone 05;
dust post fired
Photo © Malcom Bennett

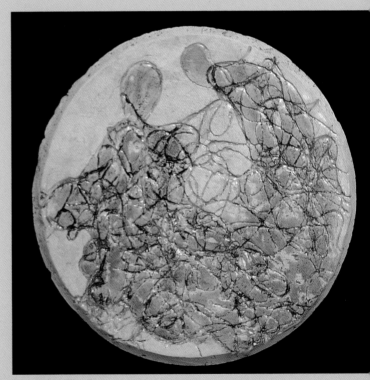

Paul Andrew Wandless
Entwined, 1994
15½ x 15½ x 3 inches (39.4 x 39.4 x 7.6 cm)
High-fire stoneware; twine relief print
Photo © artist

Gary Carlos
Target, 2003
62 x 62 inches (157.5 x 157.5 cm)
Press-molded earthenware tiles; electric fired, cone 04;
stenciled underglaze and glaze
Photo © artist

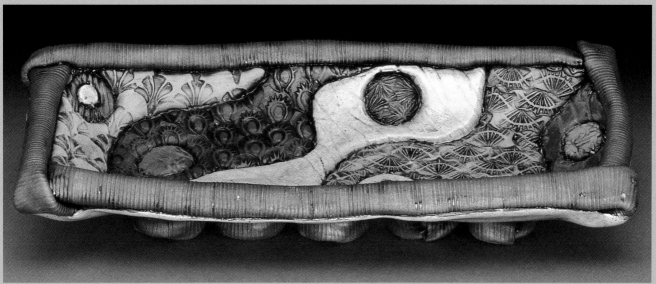

Amy Sanders
Rectangular Pieced Bowl, 2004
3½ x 18 x 8 inches (8.9 x 45.7 x 20.3 cm)
Slab-built stoneware; electric fired, cone 6; stamp impressed
Photo © Shane Baskin

David L. Gamble
From the Manhole Cover Series, 2005
11 x 13 x 1½ inches (27.9 x 33 x 3.8 cm)
Terra cotta; bisque fired, cone 03; lustre-gold glaze over
black copper oxide wash; electric fired in oxidation, cone 04
Photo © artist

David L. Gamble
From the Manhole Cover Series, 2005
11 x 13 x 1½ inches (27.9 x 33 x 3.8 cm)
Terra cotta; bisque fired, cone 03; lustre-gold glaze over
black copper oxide wash; electric fired in oxidation, cone 04
Photo © artist

Top left: **Douglas E. Gray**
Stuppa of Good Fortune, 2004
13 x 6 x 6 inches (33 x 15.2 x 15.2 cm)
Slab-built stoneware; reduction fired, cone 10;
ink transfer on green ware
Photo © artist

Top right: **Scott Rench**
Worship, 2004
13½ x 12½ x 4 inches (34.3 x 31.8 x 10.2 cm)
Brent slab roller, terra-cotta body; freehand
brush work with screen-printed underglaze,
cone 04
Photo © Smitty Image

Left: **Chris Stanley**
Judith and Lorena, 1999
30 x 30 x 4 inches (76.2 x 76.2 x 10.2 cm)
Slip-cast porcelain; hand-drawn cut paper
stencils, engobe decoration; cones 10 and 11
Photo © Eric R. Johnson

Right: **Vanessa Grubbs**
Busy Lizzy Vase, 2004
12 x 8 x 3 inches (30.5 x 20.3 x 7.6 cm)
Thrown and altered earthenware; electric fired, cone 04;
silk-screened decal, cone 018; rubber stamping
Photo © Eddie Ing

Bottom: **Shalene Valenzuela**
Conversational Hearts, 2002–2003
1 x 5½ x 2 inches (2.5 x 14 x 5.1 cm) each
Slab-constructed low-fire clay; underglaze, glaze,
cone 06; silk screened
Photo © artist

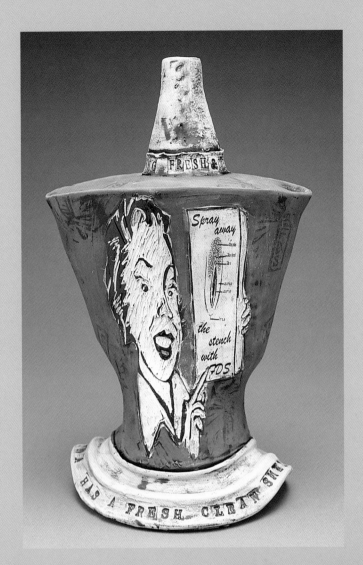

Top: **Jacqueline Kierans**
One Week, 2004
5 x 5 x 2½ inches (12.7 x 12.7 x 6.4 cm) each
Press-molded stoneware; electric fired, cone 5;
silk-screened with underglaze, in-glaze
silk-screened decal, cone 04
Photo © Gregory R. Staley

Bottom left: **Joseph Pintz**
One Dozen Milk Bottles, 2004
Crate, 12 x 15 x 10 inches (30.5 x 38.1 x 25.4
cm); bottles, each 8½ x 3 x 3 inches (21.6 x 7.6
x 7.6 cm)
Press-molded earthenware, found object; soda
fired, cone 03; screen print, cone 015
Photo © artist

Bottom right: **Marc Burleson**
Deer, Star, Gun, 2003
16 x 16 inches (40.6 x 40.6 cm)
Cast earthenware; silk-screen print,
commercial decals
Photo © artist

Slip, Engobe, and Clay Recipes for Transfer Processes

These recipes are suitable for any of the processes in this book. They're easy to use and the ingredients are widely availability. Always test the recipe before using it on your project.

Slips

White Casting Slip #1
Cone 3

Note: Dissolve the sodium silicate in hot water before adding it to the dry materials.

33 g	EPK
24 g	Ball clay
19 g	Potash feldspar
19 g	Flint
5 g	Nepheline syenite

Add
0.5 g	Sodium silicate
35 g	Water

White Casting Slip #2
Cone 5–6

Note: Dissolve the Darvan in the hot water before adding it to the dry materials.

15 g	Grolleg kaolin
15 g	Tile #6 clay
3.5 g	EPK
10 g	OM-4 ball clay
20 g	Flint
1.5 g	Ferro frit 3124
5 g	Pyrophyllite
30 g	Nepheline syenite

Add
35 g	Water
0.35 g	Darvan #7

Anna's Casting Slip
Cone 10

Note: Dissolve the Darvan and soda ash in hot water before adding them to the dry materials.

8 g	EPK
7 g	Velvacast kaolin
35 g	Grolleg
20 g	Silica, 200 mesh
30 g	Custer feldspar

Add
40 g	Water
0.10 g	Soda ash
0.35 g	Darvan #811

White Slip
Cone 10

Note: This slip can be deflocculated and used to make clay slabs for any of the image transfer processes. To use the recipe for ceramic ink printing, color it by adding 5 to 15 percent of a commercial stain. Refer to the stain manufacturer's directions for special recommendations regarding individual colors. To achieve the best color results, most stains should be oxidation fired.

35 g	OM-4 ball clay
20 g	EPK
15 g	Nepheline syenite
25 g	Silica
5 g	Whiting

Engobes

White Engobe Base
Cone 9–10

Note: Color this base recipe with commercial stains in amounts of 5 to 15 percent. Read the stain manufacturer's directions for special recommendations for individual colors. For the best color results, most stains are meant to be oxidation fired.

19 g	OM-4 ball clay
18 g	EPK
23 g	Nepheline syenite
27 g	Silica
4 g	Borax
8 g	Zircopax

Red Engobe
Cone 04–2

68 g	Redart
10 g	Ball clay
10 g	Nepheline syenite
12 g	Lithium carbonate

Clay Bodies

White Earthenware
Cone 04

38 g	Kentucky ball clay
57 g	Talc
5 g	EPK

Red Sculpture Clay
Cone 04

42 g	Redart clay
14 g	Talc
14 g	Cedar Heights Goldart
14 g	A. P. Green Fire
14 g	OM-4 ball clay
2 g	Wollastonite

Add

7 g	Grog

Tile Clay Porcelain
Cone 10–11

39 g	#6 Tile clay
19 g	OM-4 ball clay
19 g	Flint
23 g	Potash feldspar

Cone Charts

Temperature Equivalents °F for Orton Pyrometric Cones

		Self Supporting Cones 1¾" mounting height						Large Cones				Small 15⁄16" height	
		Regular			Iron Free			Regular		Iron Free		Regular	PCE
		27	108	270	27	108	270	108	270	108	270	540	270
Soft Series	022	1049	1087	1094				N/A	N/A			1166	
	021	1076	1112	1143				N/A	N/A			1189	
	020	1125	1159	1180				N/A	N/A			1231	
	019	1213	1252	1283				1249	1279			1333	
	018	1267	1319	1353				1314	1350			1386	
	017	1301	1360	1405				1357	1402			1443	
	016	1368	1422	1465				1416	1461			1517	
	015	1382	1456	1504				1450	1501			1549	
	014	1395	1485	1540				1485	1537			1598	
	013	1485	1539	1582				1539	1578			1616	
	012	1549	1582	1620				1576	1616			1652	
	011	1575	1607	1641				1603	1638			1679	
Low Temp Series	010	1636	1657	1679	1600	1627	1639	1648	1675	1623	1636	1686	
	09	1665	1688	1706	1650	1686	1702	1683	1702	1683	1699	1751	
	08	1692	1728	1753	1695	1735	1755	1728	1749	1733	1751	1801	
	07	1764	1789	1809	1747	1780	1800	1783	1805	1778	1796	1846	
	06	1798	1828	1855	1776	1816	1828	1823	1852	1816	1825	1873	
	05½	1839	1859	1877	1814	1854	1870	1854	1873	1852	1868	1909	
	05	1870	1888	1911	1855	1899	1915	1886	1915	1890	1911	1944	
	04	1915	1945	1971	1909	1942	1956	1940	1958	1940	1953	2008	
	03	1960	1987	2019	1951	1990	1999	1987	2014	1989	1996	2068	
	02	1972	2016	2052	1983	2021	2039	2014	2048	2016	2035	2098	
	01	1999	2046	2080	2014	2053	2073	2043	2079	2052	2070	2152	
Intermediate Series	1	2028	2079	2109	2046	2082	2098	2077	2109	2079	2095	2163	
	2	2034	2088	2127				2088	2124			2174	
	3	2039	2106	2138	2066	2109	2124	2106	2134	2104	2120	2185	
	4	2086	2124	2161				2120	2158			2208	
	5	2118	2167	2205				2163	2201			2230	
	5½	2133	2197	2237				N/A	N/A			N/A	
	6	2165	2232	2269				2228	2266			2291	
	7	2194	2262	2295				2259	2291			2307	
	8	2212	2280	2320				2277	2316			2372	
	9	2235	2300	2336				2295	2332			2403	
	10	2284	2345	2381				2340	2377			2426	
	11	2322	2361	2399				2359	2394			2437	
	12	2345	2383	2419				2379	2419			2471	2439

Temperatures shown are for specific mounted height above base. For Self-Supporting - 1¾". For Large - 2". For Small Cones - 15⁄16".
For Large Cones mounted at 1¾" height, use Self-Supporting temperatures. The actual bending temperature depends on firing conditions.
Charts provided with permission from The Edward Orton Jr. Ceramic Foundation.

Temperature Equivalents °C for Orton Pyrometric Cones

| Cone | Self Supporting Cones 1¾" mounting height | | | | | | Large Cones | | | | Small 15/16" height | |
| | Regular | | | Iron Free | | | Regular | | Iron Free | | Regular | PCE |
	15	60	150	15	60	150	60	150	60	150	300	150
Soft Series												
022	565	586	590				N/A	N/A			630	
021	580	600	617				N/A	N/A			643	
020	607	626	638				N/A	N/A			666	
019	656	678	695				676	693			723	
018	686	715	734				712	732			752	
017	705	738	763				736	761			784	
016	742	772	796				769	794			825	
015	750	791	818				788	816			843	
014	757	807	838				807	836			870	
013	807	837	861				837	859			880	
012	843	861	882				858	880			900	
011	857	875	894				873	892			915	
Low Temp Series												
010	891	903	915	871	886	893	898	913	884	891	919	
09	907	920	930	899	919	928	917	928	917	926	955	
08	922	942	956	924	946	957	942	954	945	955	983	
07	962	976	987	953	971	982	973	985	970	980	1008	
06	981	998	1013	969	991	998	995	1011	991	996	1023	
05½	1004	1015	1025	990	1012	1021	1012	1023	1011	1020	1043	
05	1021	1031	1044	1013	1037	1046	1030	1046	1032	1044	1062	
04	1046	1063	1077	1043	1061	1069	1060	1070	1060	1067	1098	
03	1071	1086	1104	1066	1088	1093	1086	1101	1087	1091	1131	
02	1078	1102	1122	1084	1105	1115	1101	1120	1102	1113	1148	
01	1093	1119	1138	1101	1123	1134	1117	1137	1122	1132	1178	
Intermediate Series												
1	1109	1137	1154	1119	1139	1148	1136	1154	1137	1146	1184	
2	1112	1142	1164				1142	1162			1190	
3	1115	1152	1170	1130	1154	1162	1152	1168	1151	1160	1196	
4	1141	1162	1183				1160	1181			1209	
5	1159	1186	1207				1184	1205			1221	
5½	1167	1203	1225				N/A	N/A			N/A	
6	1185	1222	1243				1220	1241			1255	
7	1201	1239	1257				1237	1255			1264	
8	1211	1249	1271				1247	1269			1300	
9	1224	1260	1280				1257	1278			1317	
10	1251	1285	1305				1282	1303			1330	
11	1272	1294	1315				1293	1312			1336	
12	1285	1306	1326				1304	1326			1355	1337

Temperatures shown are for specific mounted height above base. For Self-Supporting - 1¾". For Large - 2". For Small Cones - 15/16".
For Large Cones mounted at 1¾" height, use Self-Supporting temperatures. The actual bending temperature depends on firing conditions.
Charts provided with permission from The Edward Orton Jr. Ceramic Foundation.

Diazo Screen Emulsion Exposure Times

Here are some suggested times for burning a screen sensitized with diazo emulsion as demonstrated in this book, but also read and follow the instructions and suggested times on the particular brand of emulsion purchased. Times may vary based on the emulsion used, the thickness of its application, the age of the emulsion, and the strength of the light source used. I used a 250-watt photoflood bulb in a 10½-inch (26.7 cm) diameter polished-metal reflector.

Screen Size	Bulb-to-Screen Distance	Exposure Time
8 x 10 inches (20.3 x 25.4)	12 inches (30.4 cm)	10–12 minutes
10 x 14 inches (25.4 x 35.6)	12 inches (30.4 cm)	10–12 minutes
12 x 18 inches (30.5 x 45.7)	15 inches (38.1 cm)	16–18 minutes
16 x 20 inches (40.6 x 50.8)	17 inches (43.2 cm)	20–22 minutes
18 x 20 inches (45.7 x 50.8)	17 inches (43.2 cm)	20–22 minutes

A Quick Guide to Mixing Good Plaster

Calculate the volume you'll need, then round the number so you can use this chart to find the weights of water and plaster.

1 Weigh out the water and plaster.

2 Without stirring, slowly pour the plaster into the water.

3 Slake for 2 minutes.

4 Agitate.

5 Pour the plaster into the mold box.

Volume cubic inches (cc³)	Water (quarts)	Plaster (pounds)
40 (655 cc³)	0.5 (473 mL)	1.40 (.66 kg)
80 (1,311 cc³)	1 (945 mL)	2.85 (1.3 kg)
160 (2,622 cc³)	2 (1.9 L)	5.7 (2.6 kg)
240 (3,933 cc³)	3 (2.8 L)	8.55 (3.9 kg)
320 (5,244 cc³)	4 (3.8 L)	11.4 (5.2 kg)
400 (6,555 cc³)	5 (4.7 L)	14.25 (6.6 kg)
480 (7,866 cc³)	6 (5.7 L)	17.1 (7.8 kg)
560 (9,177 cc³)	7 (6.6 L)	19.95 (9 kg)
640 (10,488 cc³)	8 (7.8 L)	22.8 (10.3 kg)
720 (11,799 cc³)	9 (8.5 L)	25.65 (11.6 kg)
800 (13,110 cc³)	10 (9.5 L)	28.5 (13 kg)

Acknowledgments

As with all endeavors in life, it takes support, encouragement, and perseverance to accomplish your goals. I received all this and more from my family, friends, and colleagues as I wrote this book.

First and foremost I thank my beautiful wife, Jane, and my young son, Miles, for being patient, understanding, supportive, and encouraging over the year and a half it took to finish this project. They allowed me to spend countless hours in front of my laptop typing, in my studio testing, or just doing general research. Jane and Miles are the foundation and inspiration that allow me to pursue my dreams with confidence. I thank my brother, Danny, for his continued support and for making it possible for me, many years ago, to pursue my career in the field of art.

No single individual is an expert in all the varied areas of art, and having others proofread one's work is very important. I'm no exception to this rule, and I'm grateful to those who did technical reading for the book. Richard Burkett did a lot of technical proofreading for me; his input was vital and greatly enhanced the accuracy of the information covered. Paul Scott was the first person I contacted before writing this book, and he has been supportive and encouraging the entire time. Paul also had very insightful input and direction and added to the accuracy of its content. Thanks to printmaker April Flanders for proofing the printmaking information.

Victor Spinski introduced me to decals and plaster many years ago and continues to share his knowledge with me whenever I need it. David Gamble, Tracey Scheich, and Scott Rench are artists and friends who also freely gave of their time and expertise. Last but not least I thank Suzanne Tourtillott. I agreed to write this book because I knew Suzanne would be my editor. It's with her support, encouragement, extreme patience, and amazing editing skills that this idea for a book became a reality.

About the Author

Born in Miami, Florida, in 1967, and raised in Delaware, artist, writer, and curator Paul Andrew Wandless currently lives in and has his studio in the Philadelphia, Pennsylvania area. His current body of work utilizes the figure and torso as a point of departure for his clay wall pieces, clay sculptures, oil paintings, and prints. Wandless's work is collected privately, and he lectures and gives workshops regarding his work and techniques across the U.S.

Wandless has been a Special Board Appointee to the National Council on Education for the Ceramic Arts (NCECA) since 2003 as the video editor for the demonstrating artists program. He served on the Board of Directors for NCECA as a Director-at-Large (2001–2003), the Board of Directors for The Clay Studio of Philadelphia (2003–2006), and has been on the steering committee for the Philadelphia Sculptors since 2001.

He co-authored *Alternative Kilns & Firing Techniques* (Lark, 2004) and writes articles for *Pottery Making Illustrated* magazine. As an independent curator Wandless specializes in multi-cultural exhibitions and also artists using image transfer processes. In 1999 he co-founded the multicultural website www.culturalvisions.org as a gallery and resource site for artists.

In 2002 Wandless received the Distinguished Young Alumni Award from Minnesota State University, Mankato. Wandless has taught as a visiting assistant professor since 1999 at different universities while maintaining his studio work. He holds a Master of Fine Arts degree from Arizona State University (1999), a Master of Arts degree from Minnesota State University-Mankato (1997), and a Bachelor of Fine Arts degree from the University of Delaware (1995).

Gallery Contributors

Dan Anderson
Edwardsville, Illinois, Pages 16, 51

Christina Antemann
Joseph, Oregon, Pages 15, 44

Elissa Armstrong
Lawrence, Kansas, Pages 45, 51

Antonio Manuel Lewis Belgrove
Nueva Gerona, Isla de la Juventud, Cuba,
Page 64

Scott Bennett
Birmingham, Alabama, Page 34

Bojidar Bonchev
Sofia, Bulgaria, Pages 58, 59

Jessica Broad
Baltimore, Maryland, Page 24

Bill Brown
Glasgow, Scotland, United Kingdom,
Pages 28, 33, 46

Richard Burkett
San Diego, California, Pages 13, 41

Mark Burleson
Atlanta, Georgia, Pages 18, 103

Gary Carlos
San Leandro, California, Pages 79, 98

Peter Castle
Cardiff, Wales, United Kingdom,
Pages 87, 97

George Cartlidge
Page 10

Kina Crow
Tujunga, California, Page 32

Charlie Cummings
Fort Wayne, Indiana, Page 78

Israel Davis
Grand Rapids, Michigan, Page 76

Nicole M. Dezelon
Pittsburgh, Pennsylvania, Page 20

Linda Doherty
Burnaby, British Columbia, Canada,
Page 85

Lenny Dowhie
Evansville, Indiana, Pages 36, 55

Diane Eisenbach
Seaside, California, Page 49

Frank James Fisher
Milford, Michigan, Pages 75, 95, 96

Anna Velkoff Freeman
Norfolk, Virginia, Page 81

Erin B. Furimsky
Bloomington, Illinois, Pages 39, 50

David L. Gamble
Indianapolis, Indiana, Page 99

Carol Gentithes
Seagrove, North Carolina, Page 53

Brian Gillis
Springfield, Illinois, Page 54

Juan Granados
Lubbock, Texas, Pages 11, 52

Douglas E. Gray
Florence, South Carolina, Pages 84, 100

Earline M. Green
Cedar Hill, Texas, Pages 82, 93

Vanessa Grubbs
Athens, Georgia, Pages 88, 101

Barbara Hanselman
Cherry Hill, New Jersey, Page 53

Rain Harris
Philadelphia, Pennsylvania, Pages 27, 55

Hodaka Hasebe
Pittsford, New York, Page 91

Stephen Heywood
Jacksonville, Florida, Pages 74, 96

Trine Hovden
Bergen, Norway, Pages 42, 47

Abby Huntoon
South Portland, Maine, Page 94

Steve Irvine
Wiarton, Ontario, Canada, Page 91

Jeff Irwin
San Diego, California, Pages 13, 48

Kirsten Jacobsen
St. Louis, Missouri, Page 95

Jacqueline Kierans
Falls Church, Virginia,
Pages 19, 92, 97, 102, 103

Janice Kluge
Birmingham, Alabama, Pages 17, 43

Howard Kottler
Page 11

Dalia Laučkaité-Jakimavičiené
Vilnius, Lithuania, Pages 30, 52

Heeseung Lee
Philadelphia, Pennsylvania, Page 22

Hope Veronica Clare Leighton
Page 8

Kicki Masthem
Portland, Oregon, Pages 61, 70, 73,

Kelly McKibben
Carbondale, Illinois, Page 69

Eric Mirabito
Alfred, New York, Pages 23, 55

Susan Feller Mollet
McKinney, Texas, Pages 29, 46

Sharon B. Murray
Silver Spring, Maryland, Page 12

Jennifer Mettlen Nolan
Hays, Kansas, Page 50

Joseph Pintz
Lincoln, Nebraska, Pages 77, 102

Lee Puffer
La Mesa, California, Page 54

David Ray
Caulfield South, Victoria, Australia, Page 37

Scott Rench
Chicago, Illinois, Pages 80, 95, 100

Sarah Rice
O'Connor, Australian Capital Territory,
Australia, Pages 60, 62, 71

Rebecca Robbins
Vancouver, British Columbia, Canada,
Pages 20, 49

Mel Robson
Brisbane, Queensland, Australia, Page 45

Meryl H. Ruth
Portland, Maine, Pages 76, 83, 96

Amy Sanders
Charlotte, North Carolina, Pages 89, 99

Amy M. Santoferraro
Gatlinburg, Tennessee, Pages 26, 35

Tracey Scheich
Newark, Delaware, Pages 33, 49

Michael T. Schmidt
Valdosta, Georgia, Page 21

Paul Scott
Wigton, Cumbria, United Kingdom,
Pages 22, 31, 52

Christine Julin Shanks
Elkton, Maryland, Pages 43, 54

Nan Smith
Gainesville, Florida, Page 45

Jasna Sokolovic
Vancouver, British Columbia, Canada,
Page 75

Victor Spinski
Newark, Delaware, Page 28

Chris Stanley
Odessa, Texas, Pages 87, 94, 100

Brendan Tang
Nanaimo, British Columbia, Canada,
Pages 14, 26

Ian F. Thomas
Butler, Pennsylvania, Pages 63, 65, 66, 72

Shalene Valenzuela
Oakland, California, Pages 72, 101

Paul Andrew Wandless
Upland, Pennsylvania,
Pages 6, 7, 40, 56, 57, 67, 68, 85, 86, 90, 97

Cherie Westmoreland
Pittsboro, North Carolina, Pages 25, 47

Shawn Williams
Lemoyne, Pennsylvania, Pages 38, 48

Index